SpringerBriefs in Computer Science

SpringerBriefs present concise summaries of cutting-edge research and practical applications across a wide spectrum of fields. Featuring compact volumes of 50 to 125 pages, the series covers a range of content from professional to academic.

Typical topics might include:

- A timely report of state-of-the art analytical techniques
- A bridge between new research results, as published in journal articles, and a contextual literature review
- A snapshot of a hot or emerging topic
- An in-depth case study or clinical example
- A presentation of core concepts that students must understand in order to make independent contributions

Briefs allow authors to present their ideas and readers to absorb them with minimal time investment. Briefs will be published as part of Springer's eBook collection, with millions of users worldwide. In addition, Briefs will be available for individual print and electronic purchase. Briefs are characterized by fast, global electronic dissemination, standard publishing contracts, easy-to-use manuscript preparation and formatting guidelines, and expedited production schedules. We aim for publication 8–12 weeks after acceptance. Both solicited and unsolicited manuscripts are considered for publication in this series.

**Indexing: This series is indexed in Scopus, Ei-Compendex, and zbMATH **

Susana Ortega Cisneros
Emilio Isaac Baungarten Leon
Pedro Mejia Alvarez

Integrated Circuit Design

Tape-out Process with Open-Source Tools

Susana Ortega Cisneros
CINVESTAV-Guadalajara
Zapopan, Jalisco, Mexico

Emilio Isaac Baungarten Leon
CINVESTAV-Guadalajara
Zapopan, Jalisco, Mexico

Pedro Mejia Alvarez
CINVESTAV-Guadalajara
Zapopan, Jalisco, Mexico

ISSN 2191-5768 ISSN 2191-5776 (electronic)
SpringerBriefs in Computer Science
ISBN 978-3-031-92107-0 ISBN 978-3-031-92108-7 (e-Book)
https://doi.org/10.1007/978-3-031-92108-7

© The Editor(s) (if applicable) and The Author(s), under exclusive license to Springer Nature Switzerland AG 2025

This work is subject to copyright. All rights are solely and exclusively licensed by the Publisher, whether the whole or part of the material is concerned, specifically the rights of translation, reprinting, reuse of illustrations, recitation, broadcasting, reproduction on microfilms or in any other physical way, and transmission or information storage and retrieval, electronic adaptation, computer software, or by similar or dissimilar methodology now known or hereafter developed.
The use of general descriptive names, registered names, trademarks, service marks, etc. in this publication does not imply, even in the absence of a specific statement, that such names are exempt from the relevant protective laws and regulations and therefore free for general use.
The publisher, the authors and the editors are safe to assume that the advice and information in this book are believed to be true and accurate at the date of publication. Neither the publisher nor the authors or the editors give a warranty, expressed or implied, with respect to the material contained herein or for any errors or omissions that may have been made. The publisher remains neutral with regard to jurisdictional claims in published maps and institutional affiliations.

This Springer imprint is published by the registered company Springer Nature Switzerland AG
The registered company address is: Gewerbestrasse 11, 6330 Cham, Switzerland

If disposing of this product, please recycle the paper.

Preface

The landscape of Integrated Circuit (IC) design has witnessed a transformative evolution over recent decades, driven by the relentless march of technological advancements and the increasing demand for more efficient, powerful, and compact devices. As we stand at the cusp of this new era, the emergence of open-source tools and processes has reshaped the paradigm, opening access to design capabilities that were once the exclusive domain of industry giants.

Whether you only have a basic knowledge of the Hardware Description Language (HDL) or looking to deepen your understanding of advanced concepts of IC design, this book provides a structured and comprehensive pathway through the complexities of Electronic Design Automation (EDA) tools and processes.

This book focuses on OpenLane and Caravel EDA tools, due to their current major role in the open-source IC design ecosystem. OpenLane provides a robust and flexible platform that automates the entire digital design flow from Register Transfer Level (RTL) to Graphic Data System II (GDSII), making it an ideal tool for teaching and learning the physical design process. Its integration with the Google-sponsored open-source MPW launcher program allowed multiple users to not only design, but also have the opportunity to fabricate real-world chips, providing invaluable hands-on experience.

Caravel, on the other hand, serves as an open-source System on a Chip (SoC) platform, allowing designers to integrate and test their designs in a versatile, real-world environment. It complements OpenLane by enabling users to package and validate their designs, bridging the gap between theoretical knowledge and practical implementation. Together, these tools provide a way to understand the full tape-out process in a way that is accessible to students, researchers, and professionals alike.

The selection of OpenLane and Caravel is therefore not only based on their technical capabilities but also on their potential to open IC design, offering a high-quality, no-cost solution that empowers those with limited access to proprietary tools, particularly in academic settings or resource-constrained environments.

The book is organized into seven main sections:

1. **Introduction**, we lay the foundation with an introduction to the rise of open-source tools in IC design, detailing their development, significance, and the innovative tape-out processes that have broadened access to semiconductor fabrication.
2. **Physical Design Flow**: This section delves into the physical design flow, tracing the history and evolution of EDA tools, comparing commercial and open-source options, and exploring the automated IC design, including critical steps like RTL synthesis, functional verification, and physical verification.
3. **Process Design Kit**: This section focuses on the Process Design Kit (PDK), essential for modern ASIC design, offering an in-depth examination of various types of ASICs, the role of standard-cell libraries, and the revolutionary impact of open-source PDKs such as SKY130 and GF180MCU.
4. **Introduction to OpenLane**: This section delves into OpenLane, an advanced open-source framework designed for digital IC design. It offers a structured and comprehensive pathway to the OpenLane design flow, including how to create and utilize custom flow scripts, interpret various outputs, and meet installation requirements. Additionally, this section covers the classification of digital circuits, enriched with practical examples and case studies demonstrating the application of OpenLane in real-world scenarios.
5. **Macro-Cells and RAM-Cells with OpenLane**: This section explores the hierarchical design with OpenLane creating macro-cells or using RAM-cells, covering macro-cell design considerations, core design processes, and examples of SRAM cells and its use.
6. **Exploring OpenLane through Case Studies and Exercises** This chapter offers a hands-on through the physical design flow using OpenLane, featuring practical examples and exercises aimed at optimizing system performance, power, and area. It presents five diverse projects— a Pseudo Random Generator, Double Floating-Point Unit, I2C Master/Slave module, an AES-128 Encryption, and a basic RISC Processor. By leveraging OpenLane and open-source PDKs, this chapter bridges the gap between design and physical implementation, enabling designers to take their projects from RTL to silicon more easily than ever before.
7. **Caravel**: The last section introduces Caravel, a platform for integrating your projects into a broader SoC environment. This chapter discusses the architecture, features and installation of Caravel, as well as the Chipignite program and the practical aspects of RTL to GDSII flow with Caravel, resulting in a manufacturing-ready design.

This book encapsulates some of today's most popular IC design tools with a highly active community of over 8,000 enthusiasts and hundreds of circuits built with the first open-source manufacturable PDKs. This manuscript explains in a simple way the use of these new tools, providing a solid foundation for you to create ever more complex ICs that solve today's needs.

CINVESTAV-Guadalajara, Mexico &
Universidad Autónoma de Guadalajara, Mexico
CINVESTAV-Guadalajara, Mexico
CINVESTAV-Guadalajara, Mexico

Emilio Isaac Baungarten Leon
Susana Ortega Cisneros
Pedro Mejía Alvarez

Contents

1	**Introduction**		1
	1.1 The Semiconductor Industry		1
		1.1.1 Semiconductor Business Models and Value Creation	2
		1.1.2 Major Semiconductor Producers Worldwide	3
		1.1.3 Concerns about the Semiconductor Industry	4
	1.2 The Rise of Open-Source Tools in IC Design		5
		1.2.1 The DARPA Program: A Pivotal Role in Advancing Open-Source EDA Tools	6
	1.3 Detailed Overview of Open-Source Tape-Out Processes		7
	1.4 Expanding the Scope: Beyond the Basics		8
2	**Physical Design Flow**		9
	2.1 History and Evolution of EDA Tools		9
	2.2 Commercial EDA Tools		11
	2.3 Open-Source EDA Tools		11
	2.4 Automated IC Design		12
		2.4.1 Design Entry	13
		2.4.2 RTL Synthesis	14
		2.4.3 Functional Verification	14
		2.4.4 Design for Testability	15
		2.4.5 Floorplanning	16
		2.4.6 Placement	17
		2.4.7 Clock Tree Synthesis	18
		2.4.8 Routing	20
		2.4.9 Static Timing Analysis	22
		2.4.10 Physical Verification	22
		2.4.11 Graphic Data System II	22
3	**Process Design Kit**		25
	3.1 Types of ASICs		25
		3.1.1 Full-Custom ASIC	27

		3.1.2	Semi-Custom ASIC	27
		3.1.3	Gate Array-Based ASIC	27
		3.1.4	Structured ASICs	28
	3.2	Standard-Cell Library		28
	3.3	Open-Source Process Design Kit		29
		3.3.1	SKY130 PDK	30
		3.3.2	GF180MCU PDK	31
4	**Introduction to OpenLane**			33
	4.1	OpenLane Design Flow		33
		4.1.1	Creating Custom Flow Scripts	35
		4.1.2	OpenLane Outputs	36
		4.1.3	Installation and Requirements	36
	4.2	Classification of Digital Circuits		38
		4.2.1	Combinational Circuits	39
		4.2.2	Combinational Circuit Example with OpenLane	41
		4.2.3	Sequential Circuits	43
		4.2.4	Sequential Circuits Example with OpenLane	45
5	**Macro-Cells and RAM-Cells with OpenLane**			49
	5.1	Macro-Cells		49
		5.1.1	Macro-Cell Design with OpenLane	50
		5.1.2	Macro-Cell Design Considerations	52
		5.1.3	Core Design with OpenLane	55
	5.2	SRAM Cells		61
	5.3	SKY130 SRAM Cell		64
	5.4	SRAM Macros with OpenLane, FIFO Memory Example		65
		5.4.1	SKY130 SRAM Library	65
		5.4.2	RTL SKY130 SRAM	66
		5.4.3	RTL FIFO Management Block	66
		5.4.4	FIFO Configuration Files	67
6	**Exploring OpenLane through Case Studies and Exercises**			69
	6.1	Pseudo Random Generator		69
		6.1.1	Configuration File	70
		6.1.2	Suggested Experiments	71
		6.1.3	Layout View	71
	6.2	Double-Precision Floating Point Unit		72
		6.2.1	Configuration File	73
		6.2.2	Suggested Experiments	73
		6.2.3	Layout View	74
	6.3	I2C Master/Slave		74
		6.3.1	Configuration File	76
		6.3.2	Suggested Experiments	76
		6.3.3	Layout View	77

6.4	AES-128 Encryption	78
	6.4.1 Configuration File	78
	6.4.2 Suggested Experiments	79
	6.4.3 Layout View	80
6.5	RISC-V Single Cycle	81
	6.5.1 Configuration File	81
	6.5.2 Suggested Experiments	82
	6.5.3 Layout View	83

7 Caravel 85
   7.1 Introduction to Caravel 85
        7.1.1 Caravel Architecture 87
        7.1.2 Features 88
        7.1.3 Caravel Installation 89
   7.2 Efabless 90
        7.2.1 Chipignite Program 91
        7.2.2 Multi-Project Wafer Sponsored by Google 91
        7.2.3 Collaboration with GlobalFoundries 92
        7.2.4 The Role of SkyWater and Efabless 92
   7.3 Caravel User Project Directories 92
   7.4 Repo Integration 93
   7.5 Verilog Integration 93
   7.6 GPIO Configuration 94
   7.7 Layout Integration 94
   7.8 RTL to GDSII Flow with Caravel 95
        7.8.1 Harden the User Macro 96
        7.8.2 LVS Issues with Voltage and Ground Pins 100
        7.8.3 Flatten the User Macro 102
   7.9 SoC Integration 102
        7.9.1 User Area Pins 104
        7.9.2 Logic Analyzer Pins 105
        7.9.3 GPIO Pins 105
   7.10 SoC Integration Example 107

References 108

Chapter 1
Introduction

Tape-out is a critical milestone in the semiconductor manufacturing process, marking the completion of the design phase and the transition to fabrication. This term, historically linked to the era when designs were recorded on magnetic tape, now represents the final handoff of verified Electronic Design Automation (EDA) data to a foundry for production.

The tape-out process involves several sophisticated steps, starting from initial design conception in electronic form, progressing through intricate stages of design, verification, and layout optimization, to finally generating a photomask used in silicon manufacturing. Each step must be meticulously managed to ensure that the final chip will operate as intended. This process is highly complex, involving the integration of billions of transistors on a single chip, which requires precision and expertise in multiple engineering domains.

1.1 The Semiconductor Industry

The steady increase in demand for advanced electronic equipment has compelled designers to develop increasingly complex devices. Advances in ASIC production technologies have enabled the creation of ICs with over ten million gates per chip. In 2021, the global semiconductor industry's revenue surpassed $550 billion and is projected to grow by over 80%, exceeding a trillion dollars by 2030 [1–3]. Deloitte estimates that the industry will employ more than two million people worldwide in 2021, with a need for an additional 100,000 skilled workers annually by 2030 [2].

As chip manufacturing processes have evolved, the overall cost of chip design has risen significantly due to increased complexity in design and implementation. According to the International Business Strategy Corporation (IBS), design costs for each new technology generation have increased by more than 50% since the 22nm process, encompassing expenses for EDA tools, design verification, IP cores, and tape-out [3–5]. For instance, the design cost for a 7nm process is approximately $300 million, while the cost for a 5nm process is just under $550 million, and for a 2nm process, it is around $725 million [4].

Consequently, the challenge of implementing high-performance chip upgrades based on process improvements has intensified, with the price-performance ratio also increasing. Moreover, technical limitations, such as the mask size in lithography machines, have made monolithic integration unsustainable for new processes aimed at enhancing functions and performance [5].

1.1.1 Semiconductor Business Models and Value Creation

According to Forbes, a company's economic power and political influence stem from its business model, which both drives and is driven by the creation of economic value [6]. The semiconductor industry is divided into four major segments, each representing a different business model:

- **Design:** Companies that design ICs to perform specific functions.
- **Fabrication:** Companies that physically manufacture these ICs, translating designs into silicon.
- **Assembly/Packaging/Test (APT):** Companies that package the ICs into chips, making them suitable for incorporation into products such as cell phones, automobiles, medical devices, and industrial equipment, among others.
- **Semiconductor manufacturing equipment:** Companies that produce the capital equipment used by the other segments to carry out and automate their functions.

Some companies, like Intel, the industry leader in sales and profit, have complex business models that integrate design, fabrication, and packaging functions. These companies are known as Integrated Device Manufacturers (IDMs). Initially, IC companies operated this way in the early years of the industry, but recent decades have seen a strong trend towards differentiation and specialization. IDMs have often split into a design company (a Fabless IC company) and a foundry. For example, in 2008, AMD split into a fabless IC company, retaining the AMD name, and a manufacturer, GlobalFoundries [6]. Most semiconductor companies now operate exclusively within one of the four segments of the supply chain:

- **Design companies:** These companies develop software and intellectual property but do not manufacture anything. They are known as Fabless IC companies.
- **Manufacturing companies:** Known as Foundries, these companies do not design their own chips but provide contract manufacturing for Fabless IC companies.
- **Chip packaging companies:** This segment is largely separate and more commoditized.
- **Equipment manufacturing companies:** These companies are completely separate from the other segments.

This trend towards specialization highlights the economic differences between these four functions and clarifies the power dynamics within the industry.

1.1.2 Major Semiconductor Producers Worldwide

The semiconductor industry is one of the pillars of modern technology and drives advances in fields such as computing, telecommunications, healthcare and automotive. Semiconductors are integral components of a multitude of electronic devices and enable functionalities that drive our daily lives. As demand for these sophisticated technologies increases, some countries have become world leaders in semiconductor production. The global semiconductor industry is dominated by a few key players, each of which contributes significantly to the production and innovation of these vital components [6, 7]. Here's a look at the five countries that lead the world in semiconductor production:

1. **Taiwan**: Taiwan, despite its small size and complex diplomatic status, is the world leader in semiconductor manufacturing. This dominance is largely due to Taiwan Semiconductor Manufacturing Co. (TSMC), which alone accounts for about 50% of global semiconductor production. Unlike companies such as Samsung or Intel, which produce semiconductors for their own products, TSMC operates as a foundry, manufacturing chips for various clients including Apple, AMD, Nvidia, and Qualcomm.
2. **South Korea**: South Korea's Samsung Electronics is a major global player in technology and one of the top semiconductor producers. Samsung functions as both an IDM, creating semiconductors for its own products, and as a foundry, producing chips for other companies. Additionally, SK Hynix contributes significantly to South Korea's semiconductor output. In 2021, semiconductors were South Korea's largest export category, comprising 15% of the country's total exports from its extensive network of over 70 fabrication plants.
3. **Japan**: Japan, known for its technological advancements, is home to more than 100 semiconductor fabrication facilities, operated by Japanese, American, or Taiwanese firms. The Japanese government is actively working to expand the country's semiconductor manufacturing capabilities, reflecting its commitment to maintaining a significant role in the global semiconductor market.
4. **United States**: As of 2021, the United States controlled around 12% of the world's semiconductor manufacturing capacity, a notable decrease from 37% in 1990 due to increased production capabilities in Taiwan and China. Despite this decline, the semiconductor industry remains lucrative in the US. In 2021, semiconductor exports contributed $62 billion to the US economy, ranking among the top exports. US companies, while holding only 12% of global manufacturing capacity, control about 46.3% of the semiconductor market share. This discrepancy is partly due to their overseas fabrication plants and the significant value of imported US semiconductors. The COVID-19 pandemic caused disruptions in semiconductor manufacturing and supply chains, leading the US government to focus on boosting domestic production.
5. **China**: China is a key manufacturing hub and is actively working to increase its semiconductor production capacity. As the world's largest semiconductor market, China aims to achieve self-sufficiency in semiconductor manufacturing by 2030.

The Chinese government has set a goal to produce up to 25% of the world's semiconductors domestically, reducing dependence on imports and strengthening China's position in the global semiconductor industry.

1.1.3 Concerns about the Semiconductor Industry

KPMG LLP and the Global Semiconductor Alliance (GSA) conducted the 19th annual global semiconductor industry survey in the fourth quarter of 2023. This survey captures insights from 151 semiconductor executives about their outlook for the industry in 2024 and beyond. More than half of the respondents are from companies with more than $1 billion in annual revenue.

The semiconductor industry faced numerous challenges in 2023, including inflationary pressures, geopolitical uncertainties, inventory surpluses, ongoing supply chain disruptions, demand challenges in the PC and mobile device markets, and a scarcity of skilled talent. These factors contributed to a global revenue decline of 8.2% compared to 2022. Despite these hurdles, the overall industry outlook remains strong for 2024, with expectations of double-digit revenue growth. Key trends such as generative artificial intelligence, cloud computing, data centers, the increasing number of semiconductors in automobiles, and growing aerospace and defense budgets are anticipated to help the industry navigate broader economic and geopolitical risks.

The most significant concern among semiconductor executives is talent acquisition and retention, cited as the top issue for the third consecutive year from 2021 to 2023. As a result, talent development and retention remain the number one strategic priority for the industry, with competition for talent being a major concern due to the emergence of nontraditional semiconductor players enhancing their silicon capabilities.

The top three strategic priorities for semiconductor companies are:

1. **Talent development and retention**
2. **Supply chain flexibility**
3. **Implementing Generative AI**

From a geographic perspective, US respondents highlighted that talent risk slightly falls behind the nationalization of semiconductor technology as a critical issue. This balance coincides with the U.S. goal of building new factories and establishing a supply chain less dependent on Asia. Unfortunately, the outlook for the talent supply in the US is not encouraging, with an estimated 67,000 technical, computer science, and engineering jobs potentially unfilled by 2030. Companies in the Asia/Pacific region, respondents agreed that talent risk is the top issue but also viewed high foundry costs and excess production capacity as significant concerns. European respondents, while also concerned about talent risk, emphasized territorialism and global inflation. The EU is heavily investing in initiatives like the

European Chips Skills 2030 Academy program to develop a pipeline of 500,000 microelectronics experts necessary for the success of the European Chips Act.

Acquiring the right number of capable workers is a significant vulnerability for the semiconductor industry. Companies worldwide are implementing a variety of strategies to attract and retain talent. The competitive advantage of these companies could depend heavily on their success in this area. According to the KPMG Survey 2023, the top strategies include:

1. **University Partnerships (52%)**: Collaborating with universities to access fresh talent pools and establish research connections.
2. **Reinforcing the Employee Value Proposition (47%)**: Emphasizing unique value propositions, including benefits, work culture, and growth opportunities, to attract and retain talent.
3. **Offering Remote/Hybrid Positions (45%)**: Providing flexibility by allowing employees to work remotely or in a hybrid model.
4. **Annual Bonuses (40%)**: Rewarding employees with annual bonuses based on performance.
5. **Implementing AI/Automation (38%)**: Using technology to automate tasks, enabling employees to focus on strategic work.
6. **Mentorship Programs (37%)**: Establishing initiatives to support professional development through mentorship.
7. **Workforce Retraining (37%)**: Investing in training programs to upskill existing employees.
8. **Sign-On Bonuses for New Employees (28%)**: Offering financial incentives to attract new hires.
9. **Apprenticeship Programs (27%)**: Providing hands-on training and experience for entry-level positions.
10. **Hiring from Traditionally Underrepresented Groups (20%)**: Focusing on diversity and inclusion by hiring from underrepresented backgrounds.
11. **Above-Market Raises (15%)**: Offering salary increases above industry standards.
12. **Rapid Promotions (11%)**: Accelerating career growth through timely promotions.

The semiconductor industry, while promising, faces substantial challenges in talent acquisition and retention. The success of companies in this sector will significantly depend on their ability to attract and retain skilled professionals, ensuring their competitive advantage in the global market.

1.2 The Rise of Open-Source Tools in IC Design

The use of open-source tools in IC design presents a transformative opportunity for the industry. Traditionally dominated by proprietary software, the high costs and restrictive licenses associated with these tools have limited access to smaller firms and educational institutions. Open-source tools break down these barriers, offering a platform for innovation, collaboration, and education without the burden of high costs.

Open-source EDA tools have been gaining traction due to their ability to foster a collaborative environment where developers and designers can contribute to tool improvements, share knowledge, and rapidly innovate. This collaborative model not only speeds up the development of new tools but also ensures that these tools are rigorously tested and enhanced through community involvement.

1.2.1 The DARPA Program: A Pivotal Role in Advancing Open-Source EDA Tools

The Defense Advanced Research Projects Agency (DARPA) is a research and development agency of the United States Department of Defense. Established in 1958 in response to the Soviet Union's launch of Sputnik 1, DARPA's mission is to ensure that the United States maintains a technological edge in military capabilities. Over the decades, DARPA has been at the forefront of numerous technological breakthroughs, many of which have had significant impacts beyond military applications [8].

DARPA's initiatives cover a wide range of scientific and engineering disciplines, aiming to push the boundaries of what is possible. The agency's projects often involve collaboration with universities, industry, and government partners, and they are known for their high-risk, high-reward approach [9–11].

DARPA's programs have significantly influenced the semiconductor industry and EDA. One of the key initiatives in this domain is the Electronics Resurgence Initiative (ERI). Launched to ensure continued U.S. leadership in electronics innovation, ERI focuses on developing next-generation microelectronics technologies, including advancements in semiconductor materials, circuit design, and system architectures [12].

Next-generation intelligent systems supporting Department of Defense (DoD) applications—such as artificial intelligence, autonomous vehicles, shared spectrum communication, electronic warfare, and radar—require processing efficiency that far surpasses current commercial electronics capabilities. Achieving the performance levels necessary for these DoD applications involves developing highly complex SoC platforms that leverage the most advanced IC technologies. DoD researchers and development teams often lack the resources to implement such a strategy effectively, resulting in hardware design cycles that are two to three times longer than those in commercial settings.

To address the design expertise gap and keep pace with the exponential increase in chip complexity, the Intelligent Design of Electronic Assets (IDEA) program aims to develop a general-purpose hardware compiler. This compiler is capable of translating source code or schematics to physical layout (GDSII) for SoCs, System-In-Packages, and PCBs in less than 24 hours without human intervention [13]. The program seeks to leverage advances in applied machine learning, optimization algorithms, and expert systems to enable users with no prior design expertise to complete physical designs at the most advanced technology nodes. One of the most significant and impactful projects of this initiative is OpenROAD, which serves as the foundation for the creation of a highly automated workflow tool called OpenLane.

OpenLane is an open-source EDA tool that automates the process of converting RTL designs into GDSII layouts. It's part of the OpenROAD project and utilizes a variety of tools such as OpenROAD, Yosys, Magic, Netgen, CVC, SPEF-Extractor, and KLayout. This comprehensive flow handles all ASIC implementation steps from RTL synthesis to final layout.

One of the key features of OpenLane is its support for the SKY130 and GF180 PDKs, which are widely used in the open-source hardware community. The tool has been proven reliable and has been supported by Google, which has sponsored the manufacturing of hundreds of ICs using these PDKs.

OpenLane is designed to be accessible and easy to use, making it a popular choice for both educational purposes and professional ASIC design. Its active community of over 8,000 participants worldwide continues to contribute to its development and improvement. In the following chapters you will learn about the physical design flow with the OpenLane tool.

1.3 Detailed Overview of Open-Source Tape-Out Processes

Utilizing open-source tools for the tape-out process offers significant benefits, including cost reduction, enhanced learning opportunities, and increased flexibility in design customization. However, these advantages come with their own set of challenges, such as limited technical support, slower integration of cutting-edge technologies, and variability in tool reliability and performance.

The open-source tape-out process typically includes several stages:

1. **Design Entry and Simulation**: The process begins with the design entry, often using HDL like Verilog or Very High-Speed Integrated Circuit Hardware Description Language (VHDL). Open-source simulators like Icarus Verilog and GHDL are used for simulation and functional verification.
2. **Synthesis**: Converting the HDL description into a gate-level representation is handled by tools such as Yosys, which is capable of synthesizing complex designs.
3. **Place and Route**: Tools like OpenROAD and OpenLane manage the placement of gates and routing of interconnects on the silicon layout, optimizing the physical space and electrical performance.
4. **Layout Verification**: Ensuring the layout meets specific manufacturing rules is crucial. Open-source tools such as Magic provide Design Rule Checking (DRC) and Layout Versus Schematic (LVS) functionalities.
5. **GDSII Generation**: The final layout data is compiled into a GDSII file, a standard format for sending designs to fabrication.

Each of these steps requires careful consideration and expertise to ensure that the final product is both functional and manufacturable.

1.4 Expanding the Scope: Beyond the Basics

As the semiconductor industry continues to evolve, the role of open-source tools will likely expand, driven by the increasing complexity of designs and the growing need for more customized solutions. This book aims to provide not only an introduction to basic concepts but also an exploration of advanced topics in the use of open-source tools for tape-out. We will delve into case studies, practical challenges, and innovative solutions that have emerged from the community.

This introduction sets the stage for a detailed exploration of each stage of the tape-out process using open-source tools. As we progress through the book, we will provide in-depth discussions on each tool, practical guidance on overcoming common challenges, and insights into leveraging open-source software to achieve professional-grade IC designs.

Chapter 2
Physical Design Flow

Physical design refers to the process of translating a digital circuit design into a physical layout, detailing the placement and routing of components on a semiconductor chip to meet performance, power, and area constraints. The chapter on Physical Design Flow begins by providing a foundational understanding of EDA tools. It starts with an exploration of the history and evolution of EDA tools, highlighting the transition from manual design techniques to the advanced software solutions that have revolutionized the semiconductor industry. First section covers key innovations and milestones that have significantly improved design efficiency and accuracy. Following this, there is an introduction to the various EDA tools available today, discussing their essential role in automating and optimizing the physical design process.

The subsequent sections delve into the detailed steps involved in the physical design flow. Starting with design entry, designers use HDL to formalize their ideas, which then move into RTL synthesis, converting the design into a gate-level netlist. Functional verification ensures the design operates correctly, while Design for Testability (DFT) incorporates features to facilitate defect testing. Floorplanning and placement strategically organize and position standard cell components to meet performance goals. Clock Tree Synthesis (CTS) focuses on distributing the clock signal efficiently, and routing connects components as per the design. Static Timing Analysis (STA) checks for timing requirements, physical verification ensures manufacturability, and the GDSII finalizes the layout format for fabrication.

The chapter concludes with an explanation of the GDSII, the standard format for representing the final chip design. It details the crucial role of GDSII in transferring the design to manufacturing, ensuring that the design is accurately and reliably transferred to the manufacturing process.

2.1 History and Evolution of EDA Tools

EDA for IC has a long and rich history dating back to the 1950s. In the early days of IC design, circuits were drawn by hand on paper, and the layout was created by hand on large sheets of graph paper. This process was time-consuming and error-prone, leading to a need for automation tools to aid in the design process.

The pioneering paper on Computer-Aided Design (CAD) by CRAY and KISCH in 1956 [14] opens with the observation: "The concept of computers designing other computers has garnered considerable interest over the years." The term "Computer Aided Design" gained recognition in the late 1950s, particularly through the first CAD tool for IC design, called the Sketchpad, nowadays its use in designing electronic systems is known as EDA. Sketchpad was developed at the Massachusetts Institute of Technology and was used for both digital and analog IC design [15]. Sketchpad was followed by other early CAD tools such as Simulation Program with Integrated Circuit Emphasis (SPICE) [16] and Mask Layout Graphics Interactive Editor (MAGIC) [17].

In the 1970s and 1980s, the field of EDA continued to develop rapidly, with the introduction of tools for layout, synthesis, and verification. One of the most important developments during this time was the creation of the HDL, which allowed designers to describe their circuits using a high-level language, rather than drawing them by hand [18–22]. In the 1990s, EDA became increasingly focused on automation and optimization, with the introduction of tools such as place-and-route and timing analysis. The industry also began to consolidate, with the formation of large EDA companies such as Ansys [23], Cadence [24], Synopsys [25], and Mentor Graphics [26]. With the arrival of the new millennium, EDA continued to evolve, with the introduction of new tools for power analysis, test and debug, and design for manufacturability. Today, the industry also began to shift towards open-source tools and methodologies. Recently with the creation of projects such as OpenROAD [27] and OpenLane [28].

OpenROAD and OpenLane are two notable examples of open-source EDA. OpenROAD was introduced in 2016 by the University of California, Berkeley, and focuses on creating a complete flow for digital design and implementation. It includes tools for synthesis, placement, routing, timing analysis, and physical verification [27]. OpenLane, on the other hand, was developed in 2019 by Efabless Corporation and its contributors as an automated RTL-to-GDSII flow that integrates various open-source EDA tools. It provides a user-friendly and accessible flow for creating digital designs with a high degree of automation, including RTL synthesis, placement, routing, and physical verification [28]. Both OpenROAD and OpenLane are part of a larger trend towards open-source EDA tools and methodologies in the IC design industry. This trend is driven by a desire for greater collaboration, innovation, and flexibility in the design process, as well as the increasing complexity of IC designs [29].

Currently, EDA remains a critical part of the IC design process, enabling designers to create complex circuits with a high degree of automation and accuracy. A much more detailed overview of the history of EDA can be found in [30], while a historical survey of many of the important papers from the International Conference on Computer-Aided Design is available in [31], and an in-depth explanation of the physical design flow is provided in [32].

2.2 Commercial EDA Tools

Automated IC design tools have revolutionized the way we design and manufacture ICs, enabling engineers and designers to create complex circuits with millions of transistors, something that would have been impossible just a few decades ago. The process of IC design involves several steps, such as floorplanning, placement, routing, clock tree synthesis, static timing analysis, and physical verification, all of which are highly interdependent and require careful consideration to ensure that the final design meets the required specifications. Automated IC design tools use algorithms and heuristics to optimize the design at each step, resulting in faster design cycles, improved circuit performance, and reduced manufacturing costs. In this section, we will explore some of the popular commercial automated IC design tools available in the market and how they are used in the IC design process.

- **Cadence Design Systems**: Cadence is a leading provider of electronic design automation software, hardware, and IP. Their design tools include tools for digital and analog circuit design, simulation, verification, and more.
- **Synopsys**: Synopsys is another major player in the EDA market, offering a wide range of tools for designing and verifying complex IC. Their tools cover everything from system-level design to physical implementation and verification.
- **Mentor Graphics**: Mentor Graphics provides a comprehensive suite of tools for designing and verifying electronic systems. Their tools cover everything from system-level design to physical verification, and include tools for analog and digital design, Field-Programmable Gate Array (FPGA) design, and more.
- **Ansys**: Ansys provides a suite of tools for simulating and verifying ICs and electronic systems. Their tools cover everything from system-level design to physical implementation and verification, and include tools for analog and digital design, RF design, and more.
- **Altium**: Altium is a leading provider of Printed Circuit Board (PCB) design software, with tools for schematic capture, PCB layout, and manufacturing documentation. Their tools are widely used in the electronics industry for designing and prototyping PCBs.

2.3 Open-Source EDA Tools

In the realm of EDA, open-source tools have emerged as powerful resources for designers and engineers. These tools offer a wide range of capabilities, from layout generation to synthesis and formal verification. This section provides an overview of several prominent open-source EDA tools, including OpenROAD, OpenLane, Magic, Yosys, and others. Each tool is briefly described, highlighting its primary functions and applications in the design process. Whether you are designing a simple combinational circuit or a complex sequential system, these tools offer valuable features that can enhance your design workflow.

- **OpenROAD**: An automated digital layout generation tool that aims to provide a fully autonomous, no-human-in-loop layout generation across all process nodes.
- **OpenLane**: A suite of open-source tools that automate the process of designing physical layouts of integrated circuits.
- **Magic**: A venerable VLSI layout tool, written in the 1980s at Berkeley by John Ousterhout, now famous primarily for developing the scripting language Tcl.
- **Yosys**: A framework for Verilog RTL synthesis, which can be used for synthesis and formal verification.
- **Icarus Verilog**: A Verilog simulation and synthesis tool that operates as a compiler, transforming Verilog source into a format that can be executed.
- **Qflow**: A digital synthesis flow for semicustom digital design of ASICs (not FPGAs).
- **GHDL**: An open-source simulator for the VHDL language, which allows you to compile and execute your VHDL code.
- **KiCad**: A cross-platform and open-source electronics design automation suite for printed circuit board design.

Commercial EDA tools, while powerful and feature-rich, come with significant disadvantages. They often entail high licensing costs, making them inaccessible to smaller companies, independent developers or researchers. Additionally, the proprietary nature of these tools can limit flexibility, as users are dependent on the vendor for updates, bug fixes, and support, which may not always align with their project timelines. On the other hand, open-source EDA tools offer several advantages. They are generally free to use, lowering the barrier to entry for innovation and experimentation. Open-source tools foster a collaborative environment where users can contribute to the tool's development, quickly implement custom features, and fix issues independently. This community-driven approach can lead to rapid advancements and a diverse range of functionalities tailored to specific needs.

2.4 Automated IC Design

Automated IC design is a subset of EDA that specifically focuses on automating the process of creating digital IC designs, from the RTL to the final GDSII layout. In other words, automated IC design is a specific application of EDA tools and techniques.

The automated integrated circuits design flow consists of a series of 11 steps that are designed to take a digital design from the RTL to the final GDSII layout. These steps include a range of activities, from initial design planning and simulation to physical design implementation and verification, as illustrated in Figure 2.1. The ultimate goal of this process is to create a high-quality, manufacturable design that meets specific performance, area and power requirements. By following this standardized design flow, designers can ensure that their designs are efficient, accurate, and reliable, while also minimizing design iterations and time-to-market.

2.4 Automated IC Design

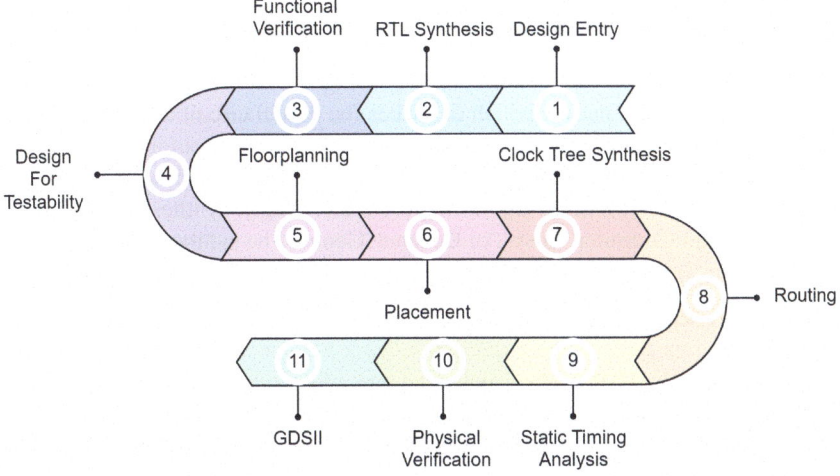

Fig. 2.1 From code to silicon: A visual guide to the RTL to GDSII flow.

2.4.1 Design Entry

Design Entry is the first step in the automated IC design flow. It involves creating a high-level representation of the design using a HDL such as Verilog or VHDL. This step requires designing the functionality of the IC based on the requirements of the customer or application. The design entry process can be performed manually or using automated tools.

During the design entry process, the IC designer creates a behavioral model of the IC using an HDL. This model represents the functionality of the IC at a high level of abstraction, without specifying the specific implementation details such as the type of logic gates, registers, or other hardware components. The designer specifies the input and output signals of the IC, along with any other necessary parameters, and creates a top-level module that encapsulates the entire design. To create the behavioral model, the designer uses coding techniques such as module instantiation, function and task calls, data types, and conditional statements. The HDL code describes how the input signals are processed and the resulting output signals are generated.

Overall, the design entry process is critical to the success of the IC design flow since it sets the foundation for the rest of the design steps. It determines the functionality, structure, and complexity of the design, and it directly impacts the design's performance, power consumption, and area. Therefore, it is essential to have a clear understanding of the design requirements and use an appropriate HDL to represent the design accurately.

2.4.2 RTL Synthesis

RTL synthesis is the process of translating a high-level HDL representation of the design into an RTL netlist, which describes the digital circuit as a collection of registers and combinatorial logic, and is then used as the starting point for subsequent steps in the design flow. During RTL synthesis, the HDL code is analyzed and transformed into an optimized digital logic circuit using a synthesis tool such as Synopsys Design Compiler, Yosys, or Cadence Genus. The synthesis tool performs a series of transformations on the HDL code to create the RTL netlist, including:

- **Syntax checking**: The HDL code is checked for syntax errors and language compliance.
- **Elaboration**: The HDL code is expanded into a hierarchical design that includes sub-modules and their connectivity.
- **Mapping**: The HDL constructs are mapped into logic gates and registers based on a library of standard cells.
- **Optimization**: The design is optimized to meet specified performance criteria such as speed, area, and power consumption. The optimization process includes techniques such as Boolean optimization, retiming, and technology mapping.
- **Timing analysis**: Timing constraints are applied to the design, and the timing of the circuit is analyzed to ensure that it meets the specified performance requirements.

The output of RTL synthesis is a netlist in a standard format, can be in the Verilog (.v) format or in the database (.ddc). This netlist is then used as the starting point for the subsequent steps in the design flow, including functional verification, design for testability, floorplanning, placement, and routing.

2.4.3 Functional Verification

Functional verification is a crucial step in the design flow that involves simulating the RTL netlist to ensure that the design functions correctly. It is an essential step in the design flow to catch design bugs early and minimize design iterations. The functional verification process involves the creation of testbenches, which are a set of test stimuli that exercise the design's functionality. The testbenches are written in HDL and are used to drive the inputs of the RTL netlist and monitor the outputs. The outputs are compared with the expected results to determine whether the design is functioning correctly.

There are two main types of simulation used in functional verification: logic simulation and timing simulation. In logic simulation, the timing delays are ignored, and the focus is on the correctness of the design's logic. In contrast, timing simulation takes into account the timing delays and is used to verify the design's timing constraints.

2.4 Automated IC Design 15

Functional verification also involves the use of Universal Verification Methodology (UVM), which is a widely used verification methodology for verifying digital designs. UVM is based on the SystemVerilog language and provides a standardized methodology for developing testbenches and testcases. UVM involves creating a testbench that models the design's environment and stimuli, and generating testcases that exercise the design's functionality. Functional verification using UVM is an iterative process that involves running simulations, analyzing the results, and making modifications to the testbench and testcases until the design meets the required functionality. Once the functional verification is complete, the design can proceed to the next step in the design flow.

2.4.4 Design for Testability

DFT is the process of designing a circuit to enable efficient testing of its functionality. The aim of DFT is to facilitate the detection and isolation of faults during the testing phase. This step typically involves adding additional circuitry to the design to enhance its testability.

One commonly used DFT technique is scan insertion, where scan chains are added to the design to enable test patterns to be applied and capture the output response, allowing for the observation of the internal state. This process involves inserting flip-flops that are chained together to form a shift register. Test patterns are shifted in serially, and the response is shifted out. Figure 2.2 illustrates the process of inserting a scan chain into a design that contains both combinational and sequential portions. The original design had three inputs and two outputs, making it difficult to initialize to a known state and observe its behavior. The addition of scan circuitry provided two additional inputs and one additional output, allowing for easier control and observation of the internal circuitry. Scan memory elements replaced the original memory elements to enable reading in scan data from the sc_in line when the sc_en line is active.

DFT is an essential step in the design flow to ensure that the design can be effectively tested and debugged. The choice of DFT technique depends on the specific design requirements and the targeted test coverage. DFT and a testbench are both used for testing the functionality of integrated circuits, but they serve different purposes. DFT is a set of techniques used during the design process to make it easier to test the finished chip. These techniques include adding extra circuitry to the design specifically for testing, such as test points, scan chains, and built-in self-test structures. The goal of DFT is to ensure that the chip can be thoroughly and efficiently tested during production, without requiring expensive and time-consuming custom test solutions. On the other hand, a testbench is a set of stimuli and expected responses used to simulate the behavior of the Device Under Test (DUT). The testbench generates input signals that are applied to the DUT and compares the resulting output to the expected behavior. The goal of a testbench is to verify that the DUT meets the desired functionality and performance specifications.

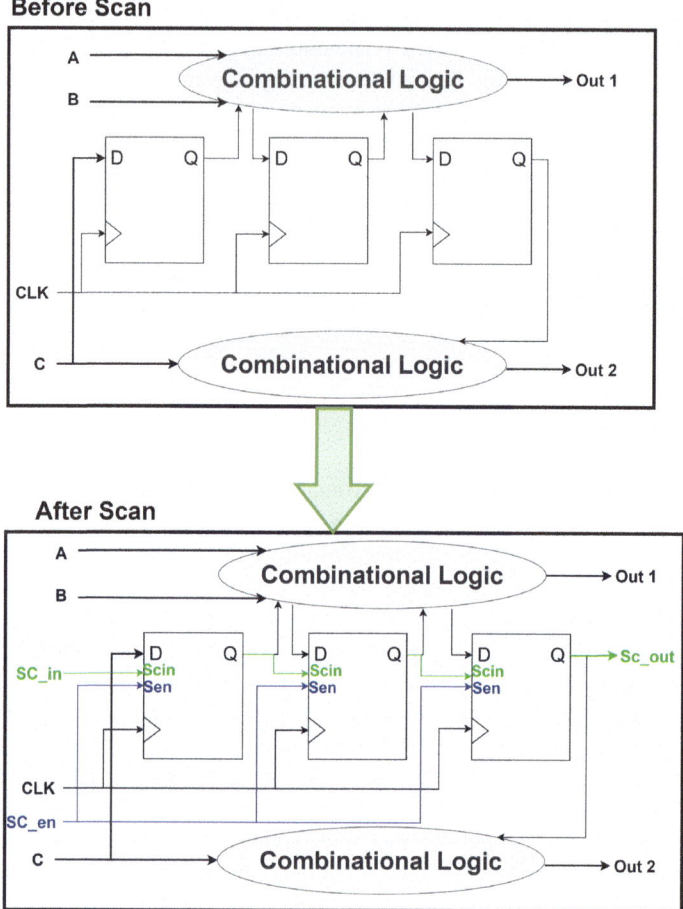

Fig. 2.2 Process of inserting a scan chain into a design.

2.4.5 Floorplanning

Floorplanning is the process of allocating chip area for different components of the design, determining the location and size of each block to ensure efficient use of space and minimize routing congestion. It is an essential step in the physical design of integrated circuits, involving the creation of a floorplan that specifies the chip area and the placement of design modules or blocks. The floorplan also includes details such as power and ground distribution, and clock distribution.

To perform floorplanning, the designer needs to consider several factors such as the size and shape of the chip, the location and size of the blocks, the interconnections between blocks, the power and signal routing, and the timing constraints. The goal

2.4 Automated IC Design

of floorplanning is to achieve an optimized design for better performance, smaller area and power consumption, higher frequency and reliability.

Figure 2.3 depicts two popular techniques employed in floorplanning: segmented floorplanning and ring and core power and ground planning. The former involves dividing the chip area into segments, each with its own set of power and ground rails, while the latter involves creating a ring around the chip with power and ground distributed throughout the core. These techniques optimize the chip's power distribution and minimize the length of power and ground lines, leading to a more efficient and reliable chip design [33].

One of the key challenges in floorplanning is minimizing the routing congestion and interconnect delay. The designer needs to consider the placement of high-speed and critical-path components to reduce the wire length and minimize the impact of parasitic capacitance and inductance. Additionally, the designer needs to ensure that the power and ground routing is adequate to meet the power and noise requirements of the design.

Floorplanning can be performed manually or using automated tools that optimize the placement of the design blocks based on predefined objectives such as area, power, or timing. Automated tools can also help the designer to evaluate different floorplan configurations and select the best one based on the specified criteria.

2.4.6 Placement

Placement is a critical step in the physical design of ICs, involving the mapping of logic cells and macros from the previous floorplanning step onto the chip area while satisfying constraints such as timing, power, and area. The goal is to find the optimal location for each cell or macro to minimize wire length and meet design requirements. Placement algorithms use mathematical models and optimization techniques to achieve this, and the process is typically divided into two main stages: global placement and detailed placement.

In global placement, the algorithm roughly places the cells on the chip area and optimizes the overall cost function. The cost function can be defined based on several objectives, such as wire length, power, timing, and area. The global placement generates a robust placement that meets the timing and congestion constraints.

In detailed placement, the algorithm refines the initial placement generated by the global placement to satisfy more detailed constraints such as timing closure and signal integrity. The detailed placement process involves several optimization techniques, such as iterative improvement, simulated annealing, and genetic algorithms. The output of the detailed placement step is the final placement of cells on the chip area that satisfies all the design constraints.

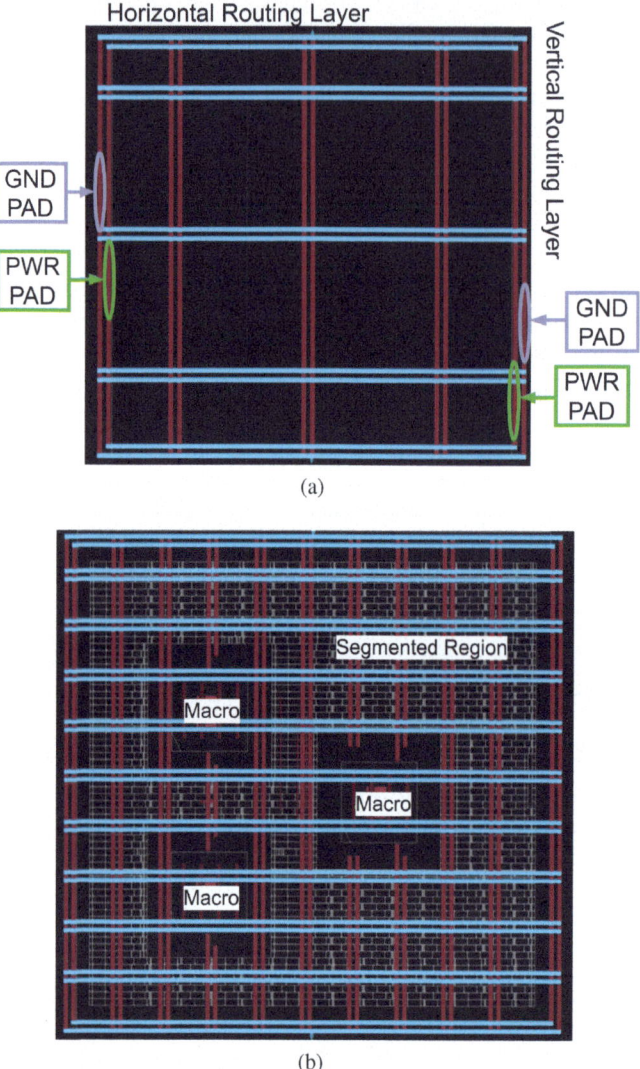

Fig. 2.3 (a) shows the power distribution network of a circuit during floor planning, highlighting the location and routing of power and ground connections, while (b) presents a Segmented Floorplan technique that divides the chip into smaller regions, facilitating the placement and routing of logic cells and macros to reduce congestion and improve overall performance and power consumption [33].

2.4.7 Clock Tree Synthesis

CTS is an important step in the physical design flow of IC, distributing a clock signal to all sequential elements in the chip, such as flip-flops, latches, and registers,

2.4 Automated IC Design

with minimal skew, delay, and power dissipation. The CTS process starts with the buffered clock signal, a high-fanout, low-skew version of the original clock, which is then distributed through multiple levels of buffers, inverters, and repeaters, forming a binary tree structure where each level represents a clock distribution network spanning a specific portion of the chip.

The CTS algorithm seeks to minimize clock skew—the variation in arrival times of the clock signal at different sequential elements—by balancing the load of each clock network and adjusting the placement and sizing of buffers to compensate for the delay and capacitance of the interconnects. There are two main types of CTS algorithms: analytical and iterative. Analytical algorithms use mathematical models to predict the timing behavior of the clock network and generate an optimized placement and sizing solution, while iterative algorithms use a trial-and-error approach to refine the clock tree until the timing constraints are met.

Figure 2.4 shows the top module layout of a design after the clock buffer insertion step,. The clock buffer insertion helps to balance the clock network by inserting buffers and maintaining uniform delay throughout the clock tree. In Figure 2.5, a detailed view of a clock buffer block is shown. The buffer block is responsible for buffering and amplifying the clock signal to ensure it reaches all the sequential elements in the design with minimal skew and delay.

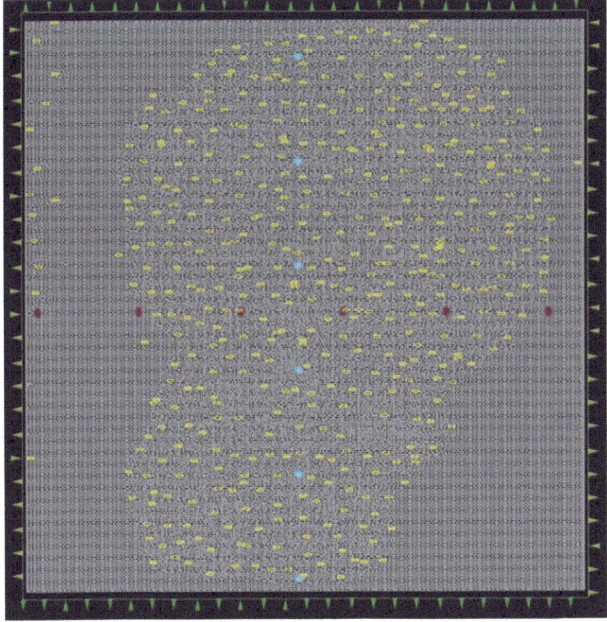

Fig. 2.4 Displays the top module layout of a design after clock buffer insertion, each yellow box represents a clock buffer cell.

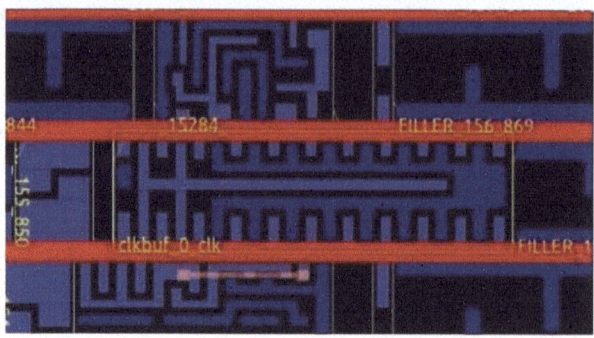

Fig. 2.5 Provides a detailed view of a single clock buffer cell. The clock tree synthesis step inserts these buffers to optimize clock distribution and reduce clock skew in the design.

2.4.8 Routing

The routing step in the physical design process involves connecting all the nets in the design using metal interconnect layers. Routing can be done in two ways: global routing and detailed routing. Global routing creates a tentative routing solution by determining the path that each net will take and the metal layers they will use. Detailed routing is the final stage, where the actual connections are made between the pins of the components using the metal layers specified by the global routing.

The routing process starts with global routing, which is done by dividing the chip into regions and determining the routes for each net using an algorithmic approach. The algorithm aims to minimize the total wire length and keep the number of vias (vertical interconnect access) and bends within specified limits. The global routing solution is then fed into the detailed routing stage, where the actual metal traces are created. Detailed routing involves the placement of wires within the routing channels, which are areas between rows of standard cells or macros, and the use of vias to connect metal layers.

Routing can be a challenging task due to the large number of nets, the limited number of metal layers, and the constraints imposed by the design rules. Congestion, where the density of wires in a particular region is too high, can occur and lead to additional design iterations. Advanced routing techniques such as multi-layer routing, differential routing, and length matching can be employed to optimize performance and reduce signal integrity issues. The final routed design is then subjected to post-layout verification to ensure that the routing meets the timing, power, and area constraints of the design.

It is important to keep in mind what each step does, the main difference between CTS and routing is that CTS involves inserting buffers to balance the clock signal distribution and optimize the clock skew as Figure 2.6 (a) shows, while routing involves making the actual physical connections between the various components of the design as Figure 2.6 (b) shows. CTS focuses on creating a stable clock signal

2.4 Automated IC Design

with low skew, while routing focuses on ensuring that all components of the design are connected properly and efficiently.

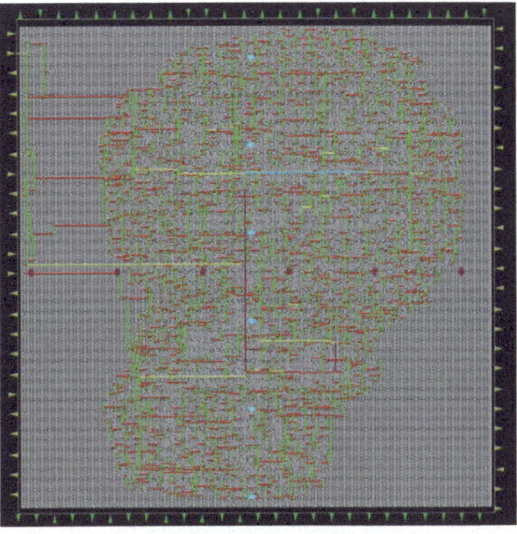

(a) Clock routing connection

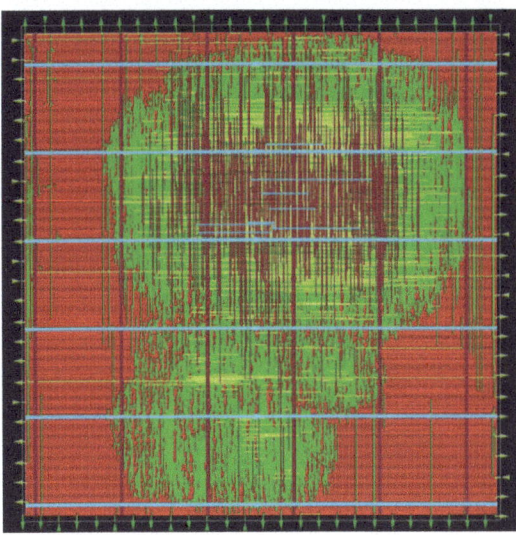

(b) Layout of a module after routing step

Fig. 2.6 (a) shows a look at the clock routing connections, which are responsible for connecting the clock tree network to the clock input pins of the flip-flops. (b) displays the layout of a module after the routing step. The routing step is responsible for making the necessary connections between the components of the design, such as wires, metal layers, and vias.

2.4.9 Static Timing Analysis

STA is a process used in the digital chip design flow to verify that the design meets the timing requirements. The objective of STA is to ensure that the signal transitions from one logic gate to another meet the timing constraints, which are usually specified by the design specifications.

STA analyzes the circuit timing by building a model of the circuit's behavior and simulating it using a timing analyzer tool. This model is created by extracting delay information from the design netlist, which is a list of the logical and physical connections between the components in the design. The tool then performs a series of simulations to analyze the critical paths in the design and measure the timing performance.

2.4.10 Physical Verification

Physical Verification is the process of ensuring that the design meets the required specifications and is free from manufacturing defects. It involves a series of checks to verify the correctness of the design layout, such as DRC, LVS checks, and parasitic extraction.

- **DRC** is a process of verifying that the layout of the design meets the manufacturing rules set by the foundry. It checks the design layout for violations such as short circuits, open circuits, narrow width, and spacing violations.
- **LVS** is a process of comparing the actual layout of the design to the original schematic. It checks whether the design layout corresponds to the intended circuit functionality and verifies that all the components of the design have been implemented correctly.
- **Parasitic Extraction** is the process of extracting parasitic elements that affect the behavior of the circuit, such as resistors, capacitors, and inductors, from the layout. The extracted parasitic elements are then used to simulate and verify the performance of the design.

Overall, the Physical Verification step is crucial in ensuring the manufacturability and functionality of the design, and it helps to identify and fix any issues that may arise during the manufacturing process.

2.4.11 Graphic Data System II

GDSII generation is the final step in the physical design process, where a file containing the design information for the chip is created. The GDSII file format, a standard readable by the foundry, includes design placement, routing information, and physical design rules, which means that the file contains all the necessary information for

2.4 Automated IC Design

manufacturing the chip, including the exact locations of each transistor, wire, and component.

The GDSII step converts the output of physical verification, including layout and design rule check information, into a GDSII file using a software tool that transforms the design database into a binary format. The file is then sent to the foundry for photomask creation and the actual manufacturing process. The GDSII file is a critical component of the physical design flow; any errors or inaccuracies can lead to significant delays or manufacturing issues. Therefore, thorough verification of the GDSII file before sending it to the foundry is essential.

Chapter 3
Process Design Kit

The objective of this chapter is to provide a comprehensive understanding of Application-Specific Integrated Circuit (ASIC) design, emphasizing its various types and the intricacies involved. We will delve into the different levels of abstraction and illustrate these concepts with practical examples to facilitate a deeper understanding of ASIC design strategies.

The structure of this chapter is as follows: First, we will explore the types of ASICs, including Full-Custom, Semi-Custom, Gate Array-Based, and Structured ASICs. Next, we will discuss the Standard-Cell Library, which is essential in ASIC design. Finally, we will examine Open-Source PDKs, focusing on the SKY130 PDK and the GF180MCU PDK, to highlight their significance and application in modern ASIC design.

3.1 Types of ASICs

The era of miniaturization spanning from 1960 to 2020 has seen significant evolutions and design changes in the field of ASICs. An ASIC design can be visualized as a small square of a few micrometers or nanometers, initially an empty box. This box is filled with functional blocks by the design team to achieve a specific functionality. The team responsible for this task is known as the front-end (logic) design team.

The backend or physical design team is responsible for tasks ranging from floor planning to physical verification at the chip level for a specific technology node. The manufacturing unit, also known as the foundry, handles the mass manufacturing and packaging of the chip. Initially, a few sample pieces are tested by the design houses to verify the intended design outcome. This task division can be found in Figure 3.1.

Skilled chip designers handle design-related tasks in various areas of chip design using EDA tools. Understanding constraints such as area, speed, and power is crucial to chip functionality. Logic design and physical design teams focus on understanding block-level and higher-level constraints to devise effective strategies to achieve the desired performance. The different types of ASICs include:

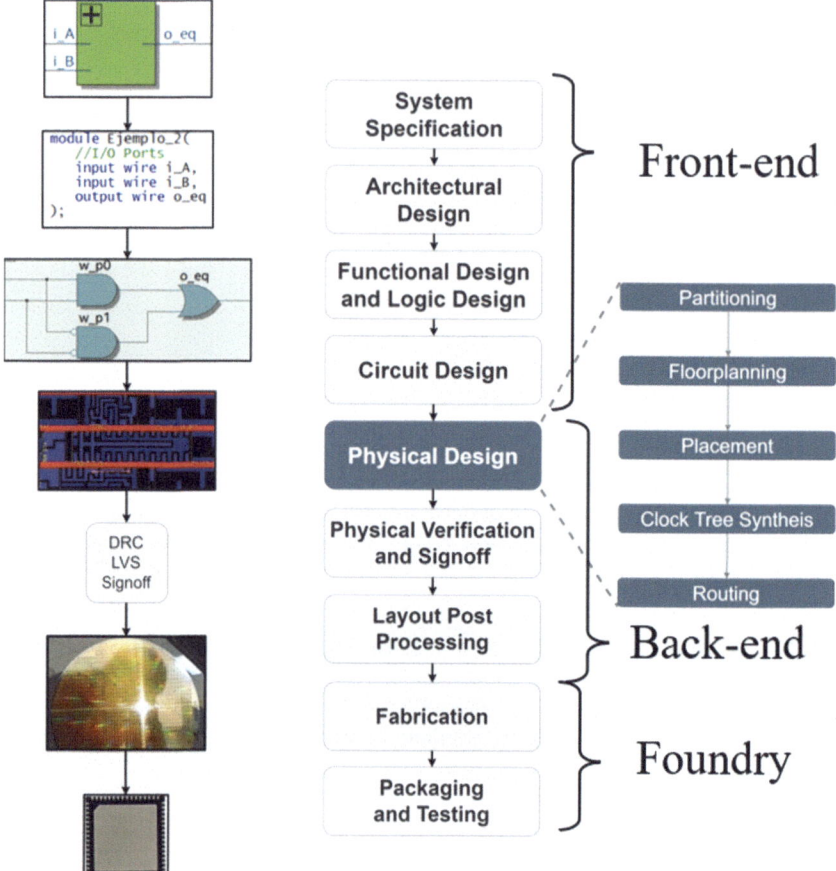

Fig. 3.1 The design flow of digital integrated circuits with Front-end, Back-end and Foundry divisions.

- **Full-custom ASIC**
- **Semi-custom ASIC**
- **Gate array-based designs**
- **Structured ASICs**

It's crucial to note that this book focuses on the design and implementation of semi-custom ASICs.

3.1 Types of ASICs

3.1.1 Full-Custom ASIC

Full-Custom ASICs are designed from scratch for a specific technology node. Each cell is tailored to meet the requirements of the technology node. This approach is beneficial for high-volume production, such as microprocessors and floating-point processors, which can be designed using the full-custom design flow. The primary advantage of the full-custom design is its ability to deliver lower power consumption, high speed, and minimal gate count for high-volume production. Although achieving speed, area, and power constraints can be time-consuming, the desired constraints can be met as the cells are designed from scratch for the desired technology nodes. The main disadvantage of this approach is the high non-recurring expenditure and the extended design cycle time.

3.1.2 Semi-Custom ASIC

Semi-custom ASIC utilize standard library cells, such as NAND, NOR, XOR, and flip-flops, during the design process. This approach leverages predefined and pre-fabricated cells, such as RAM hard macro-cores. The transistors and interconnects are customized, meaning all mask layers are customized. Compared to full-custom ASIC, the design cycle time is shorter, and pre-validated standard cells like microprocessors and macros are available during the design for a specific technology node. The disadvantage is that compared to gate array-based ASICs, the design has a high non-recurring expenditure and requires a separate fabrication mask for each design.

3.1.3 Gate Array-Based ASIC

Gate Array-Based ASICs use prefabricated wafers with an unconnected gate array, meaning wafers are common for all designs. There are mainly two types of gate array-based ASICs: Channeled Gate Array and Channel-Less Gate Array. In a Channeled Gate Array, the interconnects use the predefined spaces between the rows of base cells. In a Channel-Less Gate Array, a few top mask layers are customized. The major advantage of the gate array-based ASIC is the lower NRE cost as the same wafer is fabricated for multiple designs. Another significant advantage is the low turnaround time. The main disadvantages are the low density, lower volume, and less optimized design.

3.1.4 Structured ASICs

A Structured ASIC is an intermediate technology between a gate array and a standard cell-based ASIC. The main design task involves mapping the design into a library of building block cells and interconnecting them as necessary. The components are "almost" connected in a variety of predefined configurations, and only a few metal layers are needed for fabrication, which drastically reduces turnaround time. The advantages of the structured ASIC are low non-recurring expenditure cost, less complexity, low power consumption, high performance, and shorter marketing time. The main disadvantage is that the team needs to have a better understanding of the design constraints due to the use of prefabricated design cells.

3.2 Standard-Cell Library

A Standard-Cell Library is a collection of well-defined and pre-characterized logic cells. These cells, which include multi-drive strength and multi-threshold voltage cells, are presented in a predefined standard cell layout. The library also contains numerous physical-only cells and a set of library files required by the Place and Route (PnR).

Before a standard cell is included in the standard cell library, it undergoes a series of processes. These include schematic design, simulations, symbol creation, layout design (following standard cell layout rules), physical verifications, abstraction, extraction, and characterization. As a result, the cells in the standard cell library are free from any DRC violations, well-characterized, and suitable for the PnR tool for automatic placement and routing.

A standard cell library typically includes basic and universal gates (such as AND, OR, NOT, NAND, NOR, XOR), complex gates (like MUX, HA, FA, Comparators, AOI, OAI), clock tree cells (including Clock buffers, clock inverters, ICG cells), flip flops, latches, delay cells, physical-only cells, and scannable flip flops. Additionally, the Standard Cell Library provides files needed for auto place and route, such as LIB files (.lib), LEF files (.lef), Netlist file (.v), GDS file (.gds), SPICE Netlist (.sp), and Model file (.m).

A low drive strength cell requires less power and area but has more delay and more transition time. In contrast, a high drive strength cell can drive a larger number of cells and has a fast transition. Therefore, a PnR design engineer chooses the drive strength of cells to optimize the area, power, and performance as per the requirement.

A low threshold voltage (LVT) cell will have a lesser delay but higher leakage power compared to a high threshold voltage (HVT) cell. So, as per the requirement of timing and power, a PnR engineer uses HVT and LVT cells to balance the power and timing of the design. There is no difference in the area of multi-Vt cells. A modern standard cell library generally contains ULVT, LVT, SVT, HVT types of cells, where Vt is in increasing order.

In physical design, a variety of standard cells are added to mitigate various effects and manufacturing issues. These cells do not have any logical functions. For example, to overcome the latch-up issue, well tap cells are added. Decap cells, endcap cells, antenna cells, and filler cells are examples of such cells.

3.3 Open-Source Process Design Kit

A PDK consists of files used in the semiconductor industry to model fabrication processes for designing semi-custom ASICs. Created by the foundry, the PDK defines specific technology variations. It is then provided to customers for use in the design process. The PDK includes a library of basic photonic components, which serve as the building blocks for chip design. Designers leverage these components to create a variety of ICs, with technical and geometric representations provided in the PDK

The data in the PDK is specific to the foundry's process variation and is chosen early in the design process, influenced by the market requirements for the chip. An accurate PDK will increase the chances of first-pass successful silicon. Different tools in the design flow have different input formats for the PDK data. The PDK engineers have to decide which tools they will support in the design flows and create the libraries and rule sets which support those flows.

A typical PDK includes a primitive device library, symbols, device parameters, and PCells. It also contains verification checks, which encompass Design Rule Checking, Layout Versus Schematic checks, Antenna and Electrical rule checks, and Physical Extraction. The PDK provides technology data, including details about layers, layer names, layer/purpose pairs, colors, fills, and display attributes. It also outlines process constraints, electrical rules, and rule files in LEF and other tool-dependent rule formats.

Simulation models of primitive devices, such as transistors, capacitors, resistors, and inductors, are typically provided in SPICE or SPICE derivative formats. A Design Rule Manual is also part of the PDK, offering a user-friendly representation of the process requirements. In addition to these components, a PDK may also include standard cell libraries from the foundry, a library vendor, or those developed internally. These libraries are accompanied by the LEF format of abstracted layout data, symbols, Library (.lib) files, and GDSII layout data. All these elements together make the PDK a crucial tool in the design and development of IC.

Open-source PDKs offer numerous advantages that make them appealing for many users. One of the most notable benefits is their cost efficiency. As they are free to use, they significantly reduce design and development expenses, making them especially advantageous for startups and small companies with limited budgets. Additionally, open-source PDKs provide excellent resources for education and training, allowing students and newcomers to ASIC design to learn and experiment without the high costs associated with proprietary tools.

Another major advantage is the opportunity for collaboration and innovation. Open-source PDKs encourage contributions from designers and developers worldwide, leading to rapid advancements and the development of high-quality design tools. The transparency of open-source PDKs also plays a crucial role; users have access to the source code, enabling them to understand, modify, and customize the PDK to better suit their specific needs, thereby enhancing the flexibility of the design process.

Despite these benefits, open-source PDKs do present challenges, such as the need for technical expertise to modify and manage the tools. They can reduce dependence on specific vendors, avoiding issues like vendor lock-in, and are supported by active communities that offer invaluable support and problem-solving assistance. When considering whether to use an open-source PDK, it is important to weigh these advantages against the potential challenges.

3.3.1 SKY130 PDK

The open-source SKY130 PDK is a comprehensive set of libraries, and data files that facilitate chip design using SkyWater Technology Foundry's 130nm fabrication process. It has gained popularity and wide usage in the open-source chip design community due it is the first manufacturable free-use PDK [34, 35].

The primary objective of the SKY130 open-source PDK is to promote collaboration and innovation in chip design. It achieves this by providing free and open access to the necessary tools and resources for chip design, and for designing integrated circuits using 130nm process technology. As an open-source PDK, it enables designers to collectively share and enhance the tools and technologies used in chip design.

The open-source SKY130 PDK encompasses several features and components [35]:

- **Standard cell libraries**: It offers a broad array of basic and standard cells, such as logic gates, flip-flops, latches, etc. These cells serve as fundamental building blocks to construct more complex circuits.
- **Analog component libraries**: The PDK contains a collection of analog components, such as amplifiers, comparators, converters, etc., which are utilized in the design of analog and mixed circuits.
- **Design and verification tools**: The PDK provides numerous compatible design tools and flows that enable designers to develop, simulate, and verify their designs.
- **Technical information and documentation**: The PDK includes detailed documentation, technical specifications, and user guides that assist designers in properly using the provided libraries and tools.

The open-source SKY130 PDK has stimulated the development of open-source chip designs and projects in the community. It provides an affordable and accessible option for IC design in an older but still widely used technology such as 130nm.

3.3.2 GF180MCU PDK

The GF180MCU PDK is another open-source PDK that offers an alternative for chip design, specifically utilizing 180nm technology of GlobalFoundries. Similar to the widely used open-source SKY130 PDK, the GF180MCU PDK provides a comprehensive set of tools, libraries, and data files essential for IC design.

Designed to leverage the features and capabilities of 180nm technology, the GF180MCU PDK includes libraries of standard cells and analog components. It also provides design and verification tools, along with detailed technical documentation, to facilitate the development of chip designs using this technology [36, 37].

It's important to note that the GF180MCU PDK may have its unique features and requirements that differentiate it from the SKY130 PDK. Designers interested in using the GF180MCU PDK are advised to consult the available documentation and related resources for precise details on its implementation, usage, and compatibility with existing design tools.

As with any open-source PDK, staying updated with the latest versions and revisions of the GF180MCU PDK is crucial. This is because there may be enhancements, updates, or fixes that could impact the design and fabrication of chips using this technology.

Chapter 4
Introduction to OpenLane

In a previous chapter, readers were introduced to the RTL to GDSII flow and open-source PDKs, providing them with a basic understanding of the process. This chapter offers readers an in-depth exploration of OpenLane, an open-source automated RTL to GDSII flow for digital design. It serves as a foundational guide for users at all levels, from beginners to experienced designers, seeking to understand the capabilities and functionalities of OpenLane. The first section provides a comprehensive overview of the entire design process within OpenLane, detailing each stage from RTL synthesis to physical verification. By showcasing the seamless integration of open-source EDA tools, this section demonstrates how OpenLane streamlines the design process, ensuring efficient realization of digital designs.

The subsequent sections delve into more advanced topics, such as "Creating Custom Flow Scripts" and "OpenLane Outputs". These sections empower users to tailor the OpenLane flow to their specific design requirements and interpret the various output files and reports generated throughout the design flow effectively. Additionally, the chapter covers the installation process and system requirements in the "Installation and Requirements" section, this chapter concludes with a section of practical examples for the creation of combinational and sequential systems, ensuring that readers can set up and use OpenLane on their systems smoothly.

4.1 OpenLane Design Flow

OpenLane is an automated RTL to GDSII flow that integrates various components, including OpenROAD, Yosys, Magic, Netgen, CVC, SPEF-Extractor, KLayout, and custom scripts for design exploration and optimization [28]. This flow encompasses all ASIC implementation steps from RTL to GDSII and currently supports the A and B variants of the SKY130 PDK [35] and the GF180MCU PDK [37]. By abstracting the underlying open-source utilities, OpenLane allows users to configure the entire process with a single configuration file, providing a user-friendly, automated solution for RTL synthesis, placement, routing, and physical verification. This automation enhances design quality and reduces design iterations. Its workflow is shown in Figure 4.1.

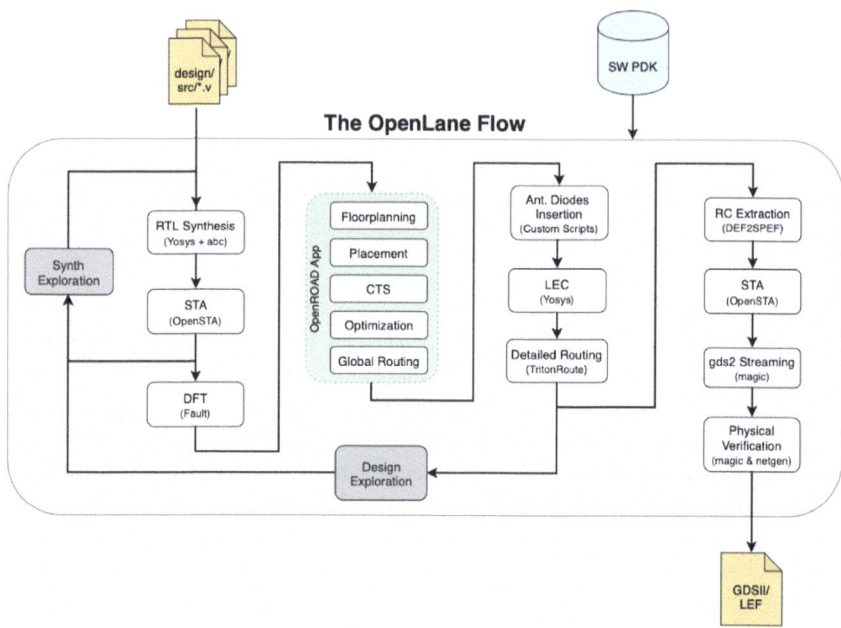

Fig. 4.1 OpenLane Architecture [28, 38].

The OpenLane flow commences with HDL synthesis, wherein the Yosys synthesis tool optimizes the design, culminating in a netlist mapped by the PDK. This phase integrates design constraints, such as clock definition and boundary conditions, and facilitates STA via the OpenSTA tool. After this, floorplanning is undertaken, utilizing OpenROAD tools for macro-related tasks, yielding a Design Exchange Format (DEF) file and delineating matrix and macro core dimensions. OpenRoad tool is utilized for chip-level floorplanning, enhancing core pin positions for optimal pad frame and core interconnect placement. Following floorplanning, standard cell, and macro placement are executed, with ensuing placement checks. The CTS phase ensues, with TritonCTS deploying clock branches integrating requisite buffers. Routing is conducted in a bifurcated approach: an initial phase with FastRoute, succeeded by a detailed process with TritonRoute. In the terminal stages, the design is subject to verifications, encompassing DRC, LVS, and STA. The successful navigation of these checks qualifies the design for approval [38–41].

Despite the apparent complexity of the OpenLane flow, it is fully automated, requiring only a single configuration file to orchestrate the entire process. This streamlines the synthesis and optimization stages, making it accessible even to newcomers to the field. With the advent of OpenLane, recent studies have conducted

4.1 OpenLane Design Flow

a comparative analysis of this open-source tool against proprietary commercial tools [38–40, 42, 43]. Table 4.1 shows some of the advantages and disadvantages of OpenLane.

Advantages of OpenLane	Disadvantages of OpenLane
Configurable through a single file, streamlining the setup	Offers less granular control compared to commercial counterparts
Fully automated, negating the need for manual intervention	Commercial tools may provide superior time optimization
Cost-free and open-source, inviting widespread use	OpenLane tends to utilize more logic cells, increasing design complexity
Simplifies the path to GDSII, reducing time and expertise required	OpenLane may lead to higher power consumption in generated designs

Table 4.1 Advantages and disadvantages of *OpenLane*.

There are several benefits to using OpenLane for digital design. One of the primary advantages is its automation, which reduces the time and effort required to complete the design process. This is particularly important for complex designs, where manual design efforts may be error-prone and time-consuming. By automating the design process, OpenLane can significantly reduce the design cycle time, leading to faster time-to-market for products. Additionally, OpenLane is open-source, meaning that it is freely available for use and can be modified and improved by users. This open-source nature encourages collaboration and innovation, allowing designers to share their work and build on the work of others.

4.1.1 Creating Custom Flow Scripts

In OpenLane, the main control script is flow.tcl, It reads a configuration file written in JavaScript Object Notation (JSON) format, which contains the design's specific parameters such as technology library, design constraints, and power management options. The configuration file can be edited by the user to customize the design flow according to their specific needs.

The configuration file acts as a navigational compass for OpenLane, delineating the sequence of script execution and defining the parameters for each stage of the flow, as elaborated in Chapter 2. For instance, the configuration file can designate the technology library for the design with the command STD_CELL_LIBRARY or set the clock frequency using CLOCK_PERIOD . A comprehensive list of such constraints is accessible at [44]. To facilitate recognition, each constraint is prefixed with a keyword indicative of its application phase within the flow, such as SYNTH for synthesis, STA for Static Timing Analysis, FP for Floorplanning, PL for placement, CTS for Clock Tree Synthesis, GRT for global routing, DRT for detailed routing, LVS for Layout vs. Schematic, among others.

Finally OpenLane is executed using the command ./flow.tcl , it reads the configuration file and runs the required Python scripts. These scripts automate the various stages of the design flow, including RTL synthesis, floorplanning, placement, clock tree synthesis, routing, static timing analysis, physical verification, and GDSII output.

4.1.2 OpenLane Outputs

By default, all output generated during the OpenLane design process is placed in a directory named after the design under the ./designs directory. For example, if the design name is my_design , the output directory would be ./designs/my_design . Within this directory, a subdirectory is created for each flow cycle, which is timestamped to allow for easy identification. Each flow cycle generates a range of output files and reports, organized into a file structure that includes directories for logs, reports, results, and intermediate files as Figure 4.2 illustrated.

The logs directory contains detailed logs of each stage of the flow, which can be useful for debugging and analysis. The reports directory contains various reports and summaries of the design, such as timing and power analysis reports. The results directory contains the final output files of the design, such as the GDSII layout file and the timing and power reports. The intermediate directory contains temporary files generated during the design process, which are useful for debugging and troubleshooting issues.

Overall, OpenLane's output directory structure provides an organized and convenient way to manage the various output files and reports generated during the design process.

4.1.3 Installation and Requirements

For installing OpenLane, it is recommended to use Ubuntu 20+. These versions of Ubuntu provide a stable environment, ensuring compatibility and smooth operation with the various third-party tools required by OpenLane. Linux distributions, particularly Ubuntu, are well supported and frequently updated, making them ideal for maintaining good tool performance at all times.

To install OpenLane, several essential packages and tools are required, including "build-essential", "python3", "python3-ven", "make", and "git". The "build-essential" package is a meta-package that includes various compilation tools necessary for building software from sources, such as the GNU Compiler Collection and other utilities. "python3" ensures that the latest version of Python is available for scripting and automation tasks within OpenLane. The "python3-venv" package

4.1 OpenLane Design Flow

allows the creation of isolated Python environments, helping manage dependencies and avoid conflicts. "make" is a build automation tool used to compile components. Lastly, "git" is a version control system essential for cloning OpenLane's source code repository, tracking changes, and collaborating with other developers. Together, these tools and packages form the foundation necessary to install and run OpenLane, ensuring a robust environment for digital design automation.

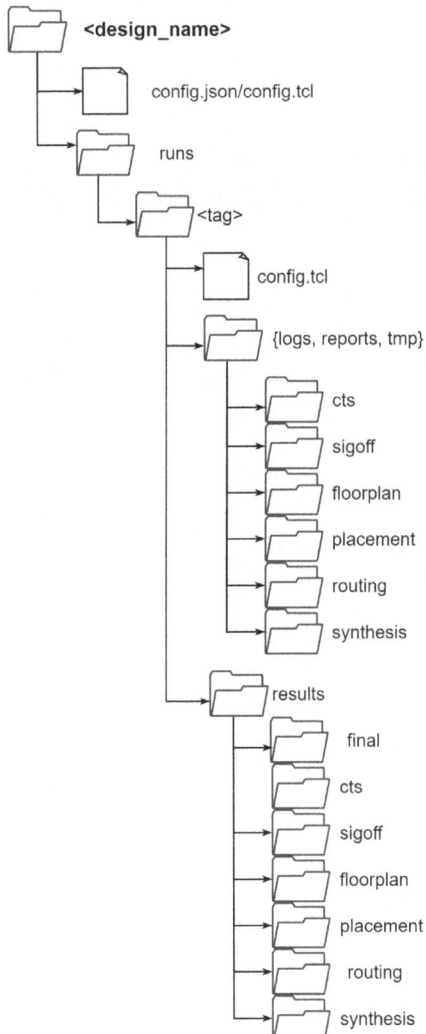

Fig. 4.2 Output files generated by OpenLane are organized into a directory structure containing logs, reports, intermediate files, and final results such as GDSII layout and timing reports.

In addition to the above, it is also necessary to install Docker. Docker is a powerful platform that simplifies the deployment of applications within software containers, providing an additional layer of abstraction and automation of virtualization. It allows developers to package an application with all of its dependencies into a standardized unit for software development. This encapsulation makes Docker an ideal tool for developers and system administrators, streamlining application development, testing, and deployment processes.

In the context of OpenLane installation, Docker offers a seamless setup experience by encapsulating OpenLane and all its dependencies within a container. This means that regardless of the host operating system, OpenLane can be run without worrying about installing the correct versions of its dependencies or configuring the environment correctly. The installation guide for Docker, which is essential for setting up OpenLane, can be found on the official Docker documentation at [45]. By following this guide, users can prepare their systems to run OpenLane with minimal hassle, ensuring a consistent and reproducible environment that mirrors production systems.

It is crucial to manage Docker as a non-root user, which is a required step in the OpenLane installation process. Neglecting this step can lead to permission issues, causing most OpenLane scripts to fail. Once the requirements are complete, the next step is to download the OpenLane repository from [44]. After downloading, navigate to the repository directory and execute the command make to start the installation process.

4.2 Classification of Digital Circuits

In the realm of digital electronics, a circuit is conceptualized as a network tasked with the processing of variables that assume discrete values. Such a circuit may be abstracted as a black box, as depicted in Figure 4.3, characterized by:

- One or more input terminals with discrete-valued signals,
- One or more output terminals also carrying discrete-valued signals,
- A functional specification that delineates the relationship between the inputs and outputs,
- A timing specification that quantifies the latency from the moment inputs alter to the subsequent response of the outputs.

Delving into the black box, we discern that circuits are an assembly of nodes and elements. An element is, in essence, a miniature circuit complete with its own inputs, outputs, and an intrinsic specification. A node, on the other hand, is represented by a wire through which a voltage conveys a discrete-valued signal [46–48].

Nodes are categorized based on their role within the circuit:

- Input nodes are the recipients of values from the external environment,
- Output nodes are responsible for transmitting values to the external environment,

4.2 Classification of Digital Circuits

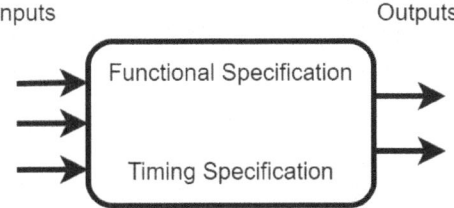

Fig. 4.3 The black box representation of a digital circuit.

- Internal nodes refer to wires that function neither as inputs nor outputs.

Digital circuits fall into two primary categories: combinational and sequential. The outputs of a combinational circuit are solely determined by the present input values, effectively synthesizing the inputs to produce the output. Logic gates exemplify combinational circuits. Conversely, sequential circuits' outputs are influenced by both the current and historical input values, indicating a dependency on the sequence of inputs. Unlike combinational circuits, which lack memory, sequential circuits incorporate memory elements to retain past input information.

4.2.1 Combinational Circuits

Combinational circuits in digital electronics are characterized by their ability to process values from the external environment through their inputs and deliver results back to them via their outputs. Internal nodes, which are not part of the inputs or outputs, serve as conduits within the system. An illustrative example is provided in Figure 4.4, showcasing a circuit with three distinct elements, labeled E1, E2, and E3, interconnected by five nodes. Nodes A, B, and C function as inputs, while nodes Y and Z serve as outputs. Node n1 is designated as an internal node, bridging elements E1 and E3.

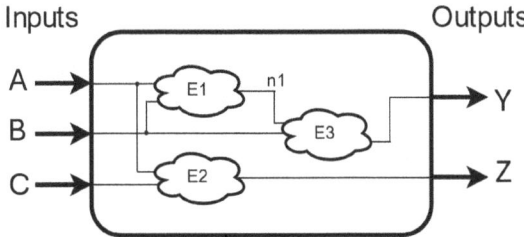

Fig. 4.4 A circuit with three elements and six nodes.

The functional specification of a combinational circuit defines the output values based on the current input values. The timing specification for such a circuit includes the minimum and maximum delays from when the inputs change to when the outputs respond.

One of the main challenges in the design of digital circuits is the optimization of timing to ensure rapid operational speed. The response time of an output to an input alteration is not instantaneous; it incurs a measurable delay. This temporal gap is crucial in circuit functionality. Figure 4.5 illustrates this concept with a timing diagram for a buffer, which visualizes the delay between an input fluctuation and the corresponding output adjustment. The diagram captures the buffer circuit's transient behavior upon an input transition.

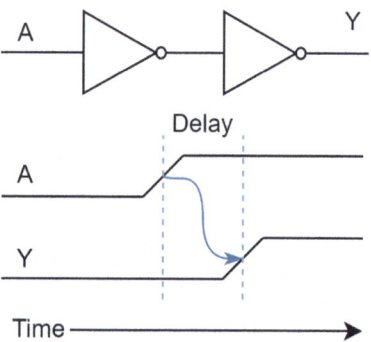

Fig. 4.5 Timing diagram depicting the delay in a buffer circuit.

The progression from a LOW to HIGH state is termed the rising edge, while the inverse transition is known as the falling edge the latter is not depicted in the figure. A notable feature in the diagram is the blue arrow, signifying that the rising edge of the output signal Y is instigated by the rising edge of the input signal A. The delay is quantified from the midpoint—specifically the 50% mark—of the input signal A to the 50% juncture of the output signal Y. This 50% point represents the signal's halfway threshold between its LOW and HIGH states during the transition.

Combinational logic systems are distinguished by two critical timing parameters: propagation delay (t_{pd}) and contamination delay (t_{cd}). The propagation delay represents the maximum duration from an input transition to the point where the outputs stabilize at their final values. Conversely, the contamination delay is the minimum duration from an input change to the initial response in any of the outputs [46, 49].

Figure 4.6 demonstrates these delays in a buffer circuit, with the propagation delay and contamination delay highlighted in blue and gray, respectively. The diagram indicates that the input A transitions from either a HIGH to LOW state or vice versa, and the output Y reacts after a certain interval. The arcs symbolize the potential start of change in Y at t_{cd} following A's transition, and the certainty that Y will reach its new state within t_{pd}.

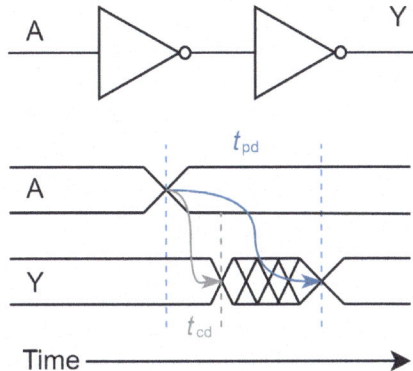

Fig. 4.6 Buffer's propagation and contamination delays.

Delays in circuits stem from factors such as the time needed to charge circuit capacitance and the finite speed of light. Variations in t_{pd} and t_{cd} arise due to:

- Disparities in rising and falling edge delays,
- The presence of multiple inputs and outputs, with some being quicker than others,
- Temperature fluctuations causing circuits to decelerate when hot and accelerate when cold.

While the computation of t_{pd} and t_{cd} delves into more granular levels of abstraction, manufacturers typically provide data sheets that detail these delays for individual gates. In our particular case all this information can be found in the official documentation of the SKY130 and GF180MCU PDK.

4.2.2 Combinational Circuit Example with OpenLane

In the previous discussion, it was established that all workflows and configurations within OpenLane are organized through a configuration file. Specifically, this is accomplished through the config.json file. When designing strictly combinational circuits, it is advisable to apply the null command within the CLOCK_PORT constraint to signify the absence of a clock signal.

Because the full adder circuit doesn't use a clock, the CLOCK_PERIOD parameter is eliminated and CLOCK_PORT is defined as a "null" value. FP_PDN_MULTI LAYER , FP_PDN_CORE_RING , and RT_MAX_LAYER are basic configuration when our module is a macro, this means that will be instantiated in another module (e.g. 8-bit adder).

FP_SIZING , DIE_AREA , and FP_CORE_UTIL are floorplan configurations, with FP_SIZING as "absolute" means that a specific size will be provided, while DIE_AREA provides the specific position of the 4-corners rectangle, "x_0 y_0 x_1 y_1", units in μm. FP_CORE_UTIL refers to the core utilization percentage and is set with a low value due to the implemented design being tiny.

The prefix PL_ means that these variables are placement configuration variables. PL_TARGET_DENSITY reflects how spread the cells would be on the core area. Due to the design being tiny (less than 100 cells) PL_RANDOM_GLB_PLACEMENT and PL_BASIC_PLACEMENT are set to 1, as well as PL_TARGET_DENSITY must have a high value.

To illustrate this, consider the example of building a full adder circuit. The configuration file for this adder would include constraints similar to the following:

```
{
    "DESIGN_NAME": "FullAdder",
    "VERILOG_FILES": "dir::src/*.v",
    "CLOCK_PORT": null,
    "FP_PDN_MULTILAYER": 0,

    "FP_PDN_CORE_RING":0,
    "RT_MAX_LAYER":"met4",
    "FP_SIZING":"absolute",
    "DIE_AREA":"0 0 30 30",
    "FP_CORE_UTIL":1,
    "PL_RANDOM_GLB_PLACEMENT":1,
    "PL_TARGET_DENSITY":0.8,
    "PL_BASIC_PLACEMENT":1
}
```

The final GDSII file is located in `<design_name>/runs/<tag>/results/final/`, as illustrated in Figure 4.2, to visualize the GDSII file you can use Klayout, Figure 4.7 show the layout result after the OpenLane execution.

OpenLane tool, as shown in the example above, offers a high degree of automation. With just a single configuration file and a single command, it can convert a Verilog file into a GDSII file. The creation of combinational circuits using OpenLane is a straightforward process. However, there are key aspects to consider during the design phase. These include the elimination of the clock signal, the area allocated for the design, and the critical path.

The critical path is particularly important as it could influence the maximum frequency if the combinational system is integrated into a more complex sequential system. In a combinational digital circuit, the critical path refers to the longest delay between an input changing value and the output changing value (the sum of t_{cd} of all the logic gates of the system, as well as their testing delay by the wire). This path imposes a limit on the maximum speed of the circuit. Therefore, careful consideration of the critical path is essential in the design of efficient digital circuits.

4.2 Classification of Digital Circuits

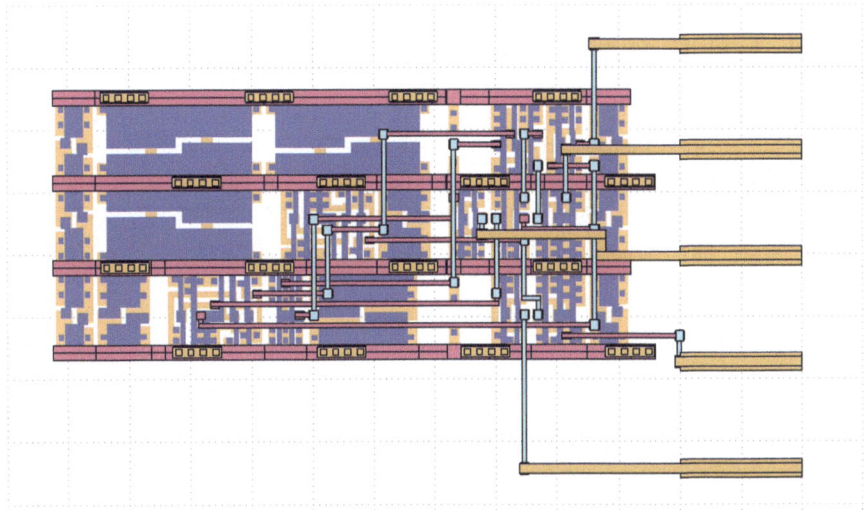

Fig. 4.7 Visualization of a full adder circuit in GDSII using Klayout.

4.2.3 Sequential Circuits

Sequential logic is characterized by its outputs, which depend on both current and prior input values, thus providing the system with memory as Figure 4.8 shown. This memory can either explicitly record certain previous inputs or condense them into a more compact form known as the system's state. The state of a sequential digital circuit is defined by a set of bits called state variables, which contain all the information about the past that is necessary to predict the circuit's future behavior.

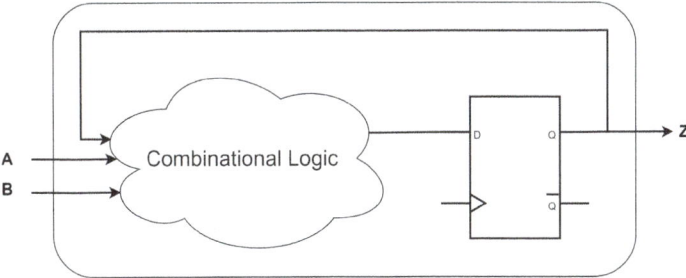

Fig. 4.8 Example of synchronous sequential circuits.

For practical purposes, we focus on building synchronous sequential circuits. These circuits are composed of combinational logic and banks of flip-flops that store the state of the circuit. A D flip-flop is a sequential element that copies the input D to the output Q on the rising edge of the clock signal. This process, known as sampling D on the clock edge, is clearly defined if D is stable at either 0 or 1 when the clock rises.

The timing of sequential elements is critical. As depicted in Figure 4.9, each element has an aperture time around the clock edge, during which the input must be stable for the flip-flop to produce a well-defined output. The aperture time is defined by the setup time (t_{setup}) and the hold time (t_{hold}), which occur before and after the clock edge, respectively. The sum of the t_{setup} and t_{hold}) times is referred to as the aperture time of the circuit, as it represents the total duration during which the input must remain stable [46, 49].

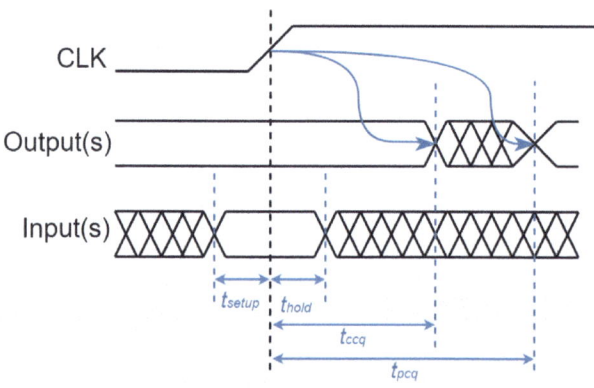

Fig. 4.9 Timing specification of a synchronous sequential circuit.

A synchronous sequential circuit, such as a flip-flop or finite state machine, also has a timing specification. When the clock signal rises, the output may begin to change after the clock-to-Q contamination delay (t_{ccq}), and must settle to its final value within the clock-to-Q propagation delay (t_{pcq}). These delays represent the fastest and slowest transmission delays through the circuit, respectively. For the circuit to sample its input correctly, the input must have stabilized at least t_{setup} before the rising edge of the clock and must remain stable for t_{hold} after the rising edge. By adhering to this dynamic discipline, we ensure the proper functioning of the flip-flops and the overall circuit.

In sequential circuits, the minimum clock period T_c is constrained by the equation:

$$T_c \geq t_{pcq} + t_{pd} + t_{setup} \tag{4.1}$$

In the realm of commercial design, the clock period is often predetermined by the Director of Engineering or the marketing department to ensure the product remains competitive. The flip-flop's t_{pcq} and t_{setup}, are provided by the manufacturer (the foundry). Consequently, we can reformulate the equation to determine the maximum allowable propagation delay through the combinational logic, which is usually the only variable in the hands of the designer:

4.2 Classification of Digital Circuits

$$t_{pd} \leq T_c - (t_{pcq} + t_{setup}) \tag{4.2}$$

The expression $t_{pcq} + t_{setup}$, known as the sequencing overhead, ideally would not encroach upon the full cycle time T_c, which is allocated for the combinational logic's computation time t_{pd}. However, this sequencing overhead inevitably reduces the available time. The aforementioned equation is referred to as the setup time constraint or the max-delay constraint. It is contingent upon the setup time and restricts the maximum delay permissible through the combinational logic.

4.2.4 Sequential Circuits Example with OpenLane

When developing a synchronous sequential circuit, careful management of the clock signal is crucial to ensure proper functionality. Uniform distribution of the clock signal minimizes skew, which can cause timing issues and incorrect circuit behavior. Choosing an appropriate clock frequency balances performance and power consumption, with higher frequencies potentially increasing heat dissipation. Timing analysis is essential for verifying that flip-flops meet setup and hold time requirements, preventing metastability and unreliable circuit operation. Additionally, accounting for propagation delays of combinational logic between flip-flops is necessary, as the maximum delay path, known as the critical path, determines the circuit's maximum operating frequency.

Understanding the primary system constraints is essential in the design of sequential circuits. These constraints typically encompass frequency, area, and power consumption. Let's consider a practical example: a 16-bit counter. This counter operates at a frequency of 50 MHz and occupies an area of less than $100um^2$. Refer to Figure 4.10 for a visual representation of this 16-bit counter. This figure provides a clear view of the counter's structure and operation. The Verilog code of the 16-bit counter is presented at [50].

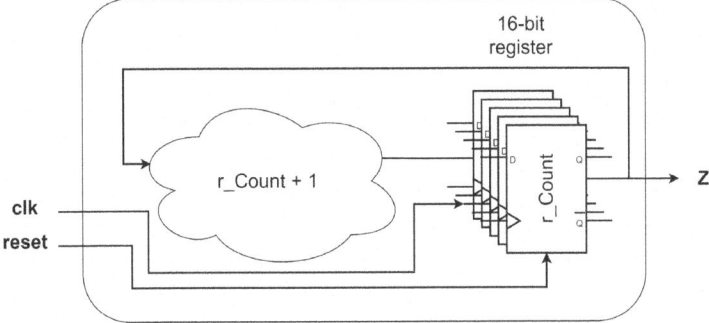

Fig. 4.10 Schematic of a 16-bit counter.

The code provided offers a distinct separation between the memory and the combinational components. The memory component is a 16-bit register, named ov_count_Q . This register undergoes modification in the first procedural block upon the detection of a positive edge of the clock or reset.

Conversely, the combinational component, implemented with the procedural block always@* , simply increments the value stored in memory by one.

When developing a sequential circuit with OpenLane, it's important to consider certain constraints such as frequency and area given by the parameters e CLOCK_PERIOD and DIE_AREA . In this case, the combinational section is relatively small, comprising less than 100 logic cells. Therefore, the configuration file for OpenLane will need to reflect these constraints to ensure optimal performance of the sequential circuit. These constraints are crucial in guiding the synthesis and place-and-route stages of the OpenLane flow, ultimately influencing the efficiency and effectiveness of the final design.

Outlined below are the parameters employed in the generation of the GDSII file for this sequential system.

```
{"DESIGN_NAME": "counter_16bit",
"VERILOG_FILES": "dir::src/*.v",
"CLOCK_PORT": "clk",
"CLOCK_PERIOD": 20.0,
"FP_PDN_MULTILAYER": true,
"FP_PDN_CORE_RING":0,
"RT_MAX_LAYER":"met4",
"FP_SIZING":"absolute",
"DIE_AREA":"0 0 80 80",
"FP_CORE_UTIL":1,
"PL_RANDOM_GLB_PLACEMENT":1,
"PL_TARGET_DENSITY":0.8,
"PL_BASIC_PLACEMENT":1 }
```

The constraints specified in the OpenLANE documentation serve various purposes in the design flow. The "DESIGN_NAME": "counter_16bit" constraint specifies the name of the design, in this case, "counter_16bit" . The "VERILOG_FILES": "dir::src/*.v" constraint points to the location of the Verilog source files for the design. The "CLOCK_PORT": "clk" constraint specifies the name of the clock signal in the design, and "CLOCK_PERIOD": 20.0 sets the target clock period for the design in nanoseconds. The "FP_PDN_MULTILAYER": true constraint enables the use of multiple metal layers for PDN generation during floorplanning. The "FP_PDN_CORE_RING":0 constraint disables the creation of a power ring around the core area during floorplanning. The "RT_MAX_LAYER":"met4" constraint sets the maximum metal layer to be used for routing. The "FP_SIZING":"absolute" constraint sets the sizing mode for floorplanning to absolute, which means the die area dimensions are specified directly. The "DIE_AREA":"0 0 80 80" constraint

4.2 Classification of Digital Circuits

specifies the coordinates of the bottom-left and top-right corners of the die area for floorplanning. The "FP_CORE_UTIL":1 constraint sets the core utilization factor during floorplanning. A value of 1 means the entire die area is used for the core. The "PL_RANDOM_GLB_PLACEMENT":1 constraint enables random global placement, which can help to avoid local minima during optimization. The "PL_TARGET_DENSITY":0.8 constraint sets the target density for placement. A value of 0.8 means the placement tool will aim to fill 80% of the available area with cells. Finally, the "PL_BASIC_PLACEMENT":1 constraint enables the basic placement mode, which performs an initial rough placement of cells. As a result, at the end of the flow, the GDSII file of this counter was obtained, which is shown in Figure 4.11.

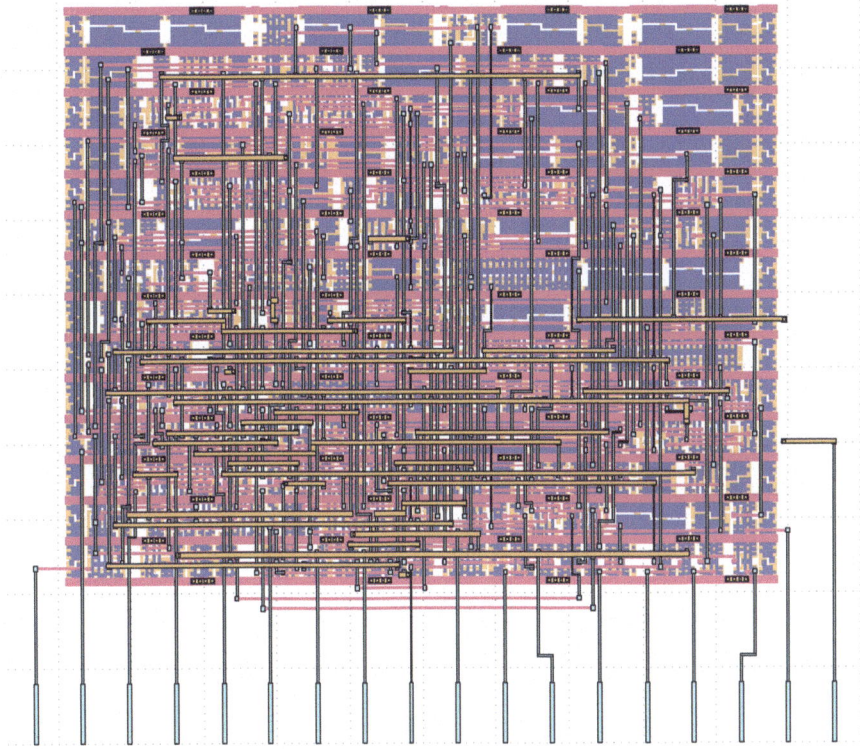

Fig. 4.11 Visualization of a 16-bit Counter circuit in GDSII using Klayout.

Chapter 5
Macro-Cells and RAM-Cells with OpenLane

In this chapter, we dive into the process of hardening designs as macros and cores using the OpenLane workflow. The objective is to familiarize readers with the steps involved in a chip-level integration using an Arithmetic Logic Unit (ALU) as an example. By following the hands-on practice outlined in this chapter, readers will gain practical experience in hardening the individual operation modules (+, -, <<, >>) and integrating them to create a fully functional ALU. Throughout the chapter, you will develop a solid understanding of the hardening process and the integration of macros and cores within the OpenLane environment.

5.1 Macro-Cells

In IC design, macro-cells are large blocks that can be considered as black-boxes. Their logic and electronic behavior are known, but their internal structural description may not always be available. As a result, macro-cells can have flexible geometries, and the pins of macro-cells can be located anywhere within the block. Macro-cells are significantly larger than standard-cells, with a typical macro-cell size being several percent of the entire placement area. Macro-cells are generally categorized into two types: soft macro-cells, which can have various specified shapes, and hard macro-cells, which have a fixed geometry [51].

Macro-cell placement is the process of positioning a given macro-cell circuit to optimize a specific objective, such as total wirelength. A complex circuit usually contains a small number of macro-cells and a large number of standard-cells. Standard-cells in the circuit have uniform height and need to be placed in specified rows, while macro-cells can be located anywhere within the layout area. Overlap between any two cells is not allowed. Macro-cell placement offers more flexibility than the classical standard-cell placement problem. In addition to determining the physical location of each cell in the netlist, the macro-cell placer also needs to determine the shapes of the soft macro-cells.

Macro-cells are used in virtually every state-of-the-art IC design. Some existing ICs are reused in the form of Intellectual Property (IP) blocks to design more complex ICs. IP blocks, embedded with various functionalities such as look-up tables, signal transforms, RAMs, etc., can be conveniently used in multimedia and communication IC designs. The reuse of design IP is considered an integral part of multi-million-gate ICs and SoC design. During the process of physical design, each IP block appears as a macro-cell. Some IP blocks may have different physical implementations, resulting in corresponding macro-cells with different geometries.

In the traditional physical design flow where the circuit contains only a few macro-cells, the circuit is first partitioned into a number of large blocks. Each block contains either one macro-cell or a group of standard-cells. Next, floorplanning is performed to determine the location and shape of each block. After floorplanning is completed, all the macro-cells are considered fixed, and a placer is used to place all the standard-cells in the circuit. Today, macro-cell placement remains a significant challenge even for commercial placement tools, and manual efforts are often needed to fix and improve the placement generated by EDA tools.

5.1.1 Macro-Cell Design with OpenLane

Designing macro-cells with OpenLane is a streamlined process, facilitated by the use of a single configuration file. When working with macro-cells, also known as Macros, certain considerations need to be taken into account. Firstly, OpenLane needs to be informed that the system being created will be a macro. Next, configurations with the PDN need to be considered to ensure that the macro will be correctly connected to the voltage and ground pins of the core (upper module). Another important aspect to consider is the routing channels and the hierarchy of the macro. It is recommended that each time you descend a level in the hierarchy, you limit the routing of the upper layer as it will be used for voltage and ground signals.

For instance, the SKY130 PDK has a total of five layers of aluminum metal. The core can use up to layer 4 for routing because layer 5 is reserved for voltage and ground signals. If a macro is implemented within that core, the routing layer should be limited to layer 4, which will primarily be used for voltage and ground tracks. Conversely, if that macro comprises other macros, these should limit their routing layer to layer 3. To illustrate this, let's consider the implementation of an ALU capable of performing additions, subtractions, and shifts to the right and left.

The corresponding Verilog codes for this system are shown below:

Addition (+) Module:

```
module Adder8Bit(input [7:0] a, b, output [7:0] sum);
    // 8-bit adder module
    assign sum = a + b;
endmodule
```

5.1 Macro-Cells

Subtraction (−) Module:

```verilog
module Subtractor8Bit(input [7:0] a, b, output [7:0] diff);
    // 8-bit subtractor module
    assign diff = a - b;
endmodule
```

Left Shift (<<) Module:

```verilog
module LeftShifter8Bit(input [7:0] a, input [2:0] b, output
↪   [7:0] result);
    // 8-bit left shifter module
    assign result = a << b;
endmodule
```

Right Shift (>>) Module:

```verilog
module RightShifter8Bit(input [7:0] a, input [2:0] b, output
↪   [7:0] result);
    // 8-bit right shifter module
    assign result = a >> b;
endmodule
```

ALU, Top Module:

```verilog
module ALU_8Bit(input [7:0] a, input [7:0]b, input [1:0]
↪   opcode, output reg [7:0] o_ALU);
    // 8-bit ALU module

    always @(*)
    begin
        case (opcode)
            2'b00: Adder8Bit add_inst(a, b, o_ALU); //
            ↪   Addition operation
            2'b01: Subtractor8Bit sub_inst(a, b, o_ALU); //
            ↪   Subtraction operation
            2'b10: LeftShifter8Bit ls_inst(a, b, o_ALU); //
            ↪   Left shift operation
            2'b11: RightShifter8Bit rs_inst(a, b, o_ALU); //
            ↪   Right shift operation
            default: o_ALU= 8'b0; // Default output
        endcase
    end
endmodule
```

Each of these operations will be performed by a specific macro, as shown in Figure 5.1.

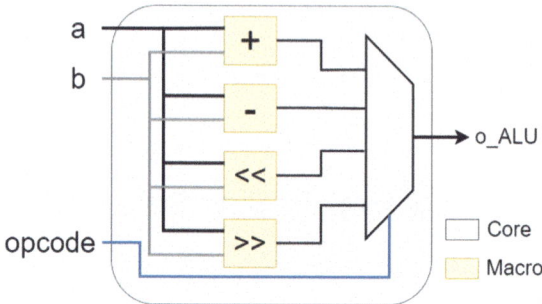

Fig. 5.1 Block diagram of an ALU implemented with macros.

Designing macro-cells with OpenLane is a straightforward process. This is largely due to the use of a single configuration file that simplifies the design process. When working with macro-cells, also known as Macros, there are certain considerations that need to be taken into account. These considerations will be explained in the following sub-sections.

5.1.2 Macro-Cell Design Considerations

The Power Grid, also known as the PDN, is a critical component in IC design that ensures proper power distribution to all chip components. It consists of a network of metal layers and power/ground rails that deliver power to various macro modules and cores. The Power Grid plays a crucial role in maintaining signal integrity and providing a stable power supply, preventing voltage drops and noise that can affect IC performance and reliability. The PDN must be carefully designed to minimize voltage drops, reduce power supply noise, and meet the power requirements of each macro module and core.

To ensure proper power delivery, the Power Grid is designed to provide an adequate power supply to all the macro modules and cores on the chip. This involves placing power/ground pads strategically, routing power lines to distribute power efficiently, and implementing decoupling capacitors to manage power fluctuations. The PDN must be carefully architected to account for the power needs of different macro modules and cores, considering their varying power consumption and spatial requirements.

In OpenLane, creating the power grid involves the crucial step of attaching specific voltage and ground pins to each module. This process is divided into two, modifications in the RTL design and adjustments in the configuration file. By carefully configuring the voltage and ground connections for each module, the power

5.1 Macro-Cells

grid can be effectively established, ensuring proper power distribution throughout the integrated circuit.

5.1.2.1 RTL Modifications

The modifications in the RTL design are required at both the macro level and core level. These changes occur at the instantiation level and the top module level. To incorporate the necessary power grid functionality, the ifdef USE_POWER_PINS parameter is added. Within this ifdef statement, the <power_pin> and <ground_pin> are included. This ensures that the power and ground connections are properly defined within the RTL code, enabling effective power distribution within the macro and core designs. For example, after the RTL changes for the Power PDN, the Addition (+) module would appear as follows:

```
module Adder8Bit(input [7:0] a, b, output [7:0] sum, `ifdef
↪ USE_POWER_PINS inout VPWR, inout VGND `endif);
    // 8-bit adder module
    assign sum = a + b;
endmodule
```

Note that in the previous example, the pins are inout, the power pin name is VPWR and the ground pin name is VGND.

All modules that will be treated as macros, including the core module, must have power and ground pins. These pins are essential for establishing the PDN within the design.

5.1.2.2 Macro Configuration File

To ensure successful integration between the core and macros, it is essential that each macro configuration file includes specific configuration variables. These variables play a crucial role in establishing a harmonious connection between the core and macros. The FP_PDN_CORE_RING variable defines the PDN ring for the core so at the macro level, it should be disabled. The RT_MAX_LAYER variable specifies the maximum routing layer allowed for the macros, ensuring compatibility with the routing resources available in the design. SYNTH_USE_PG_PINS_DEFINES This variable enables the use of power and ground pins defined within the macro. It ensures that the power and ground connections specified in the macro's Verilog code are utilized during synthesis. Additionally, the VDD_NETS and GND_NETS configuration variables identify the power and ground nets associated with the macros, enabling proper connectivity and reliable power distribution throughout the integrated circuit. By including these configuration variables, the integration process between the core and macros can be effectively managed, resulting in a cohesive and functional IC design.

The configuration variables for the Adder8Bit macro, related to power distribution and connectivity, are as follows:

```
"RT_MAX_LAYER": "met4",
"FP_PDN_CORE_RING": false,
"SYNTH_USE_PG_PINS_DEFINES": "USE_POWER_PINS",
"VDD_NETS": "VPWR",
"GND_NETS": "VGND"
```

- **RT_MAX_LAYER**: The RT_MAX_LAYER variable is set to "met4" in the Adder8Bit configuration. This restricts the maximum routing layer to metal layer 4, ensuring an efficient power grid at a core level.
- **FP_PDN_CORE_RING**: The FP_PDN_CORE_RING variable is set to "false" for the Adder8Bit macro. This indicates that a PDN ring is not required around the core module.
- **SYNTH_USE_PG_PINS_DEFINES**: The SYNTH_USE_PG_PINS_DEFINES variable is set to "USE_POWER_PINS" in the Adder8Bit configuration. This enables the usage of power and ground pins defined within the macro module during synthesis.
- **VDD_NETS**: The VDD_NETS configuration variable for the Adder8Bit macro is set to "VPWR" due to that name was previously given to it. It specifies the power supply nets or connections required for the macro module.
- **GND_NETS**: The GND_NETS configuration variable is set to "VGND" due to that name was previously given to it. It defines the ground nets or connections needed for the proper grounding within the macro module.

A similar constrain can be apply to the 4 macros due to their similarity (Adder8Bit, Subtractor8Bit, LeftShifter8Bit, and RightShifter8Bit) in terms of their inputs and die area leads to several shared configurations. These configurations ensure consistency and optimize the integration of these designs within the OpenLane workflow.

The basic configurations include DESIGN_NAME , VERILOG_FILES , CLOCK_PORT , and DESIGN_IS_CORE . These configurations define the name of the design, specify the Verilog files associated with the design, identify the clock port, and indicate whether the design is a core or a macro module.

Additional configurations such as FP_SIZING , DIE_AREA , CORE_AREA , FP_CORE_UTIL , and PL_BASIC_PLACEMENT are employed to control and optimize the area utilized by the designs. These configurations limit the footprint and utilization of the macros, ensuring efficient use of the available resources.

Furthermore, the configurations FP_PDN_HPITCH , FP_PDN_VPITCH , FP_PDN_HOFFSET , FP_PDN_VOFFSET , FP_IO_HLENGTH , and FP_IO _VLENGTH play a crucial role in mitigating PDN errors caused by the automatic scaling of voltage and ground tracks. These configurations allow for precise adjustments of pitch, offset, and length, ensuring proper power and ground connectivity throughout the designs.

5.1 Macro-Cells

The complete configuration file for the Adder8Bit macro can be found below.

```
{"DESIGN_NAME": "Adder8Bit",
"VERILOG_FILES": "dir::src/*.v",
"CLOCK_PORT": null,
"DESIGN_IS_CORE": false,
"FP_SIZING": "absolute",
"DIE_AREA": "0 0 32 32",
"CORE_AREA":"3 3 29 29",
"FP_CORE_UTIL":90,
"PL_BASIC_PLACEMENT":1,
"FP_PDN_HPITCH":10,
"FP_PDN_VPITCH":15,
"FP_PDN_HOFFSET":2,
"FP_PDN_VOFFSET":2,
"FP_IO_HLENGTH":2,
"FP_IO_VLENGTH":2,
"RT_MAX_LAYER": "met4",
"FP_PDN_CORE_RING": false,
"SYNTH_USE_PG_PINS_DEFINES": "USE_POWER_PINS",
"VDD_NETS": "VPWR",
"GND_NETS": "VGND"}
```

For more information about macro and RTL to GDSII flow visit the official documentation at [28].

5.1.3 Core Design with OpenLane

In the context of the OpenLane workflow, a chip core represents a central and fundamental component of an integrated circuit design. It serves as the foundation or main building block for the overall chip. A chip core often encompasses various other macros, which are smaller functional blocks or IP modules integrated within the core.

To configure the chip core in the OpenLane workflow, you would typically set specific environment variables in the configuration file. These environment variables define various aspects of the chip core, such as its functionality, performance, power characteristics, and interconnections with other modules. By customizing these environment variables, you can tailor the behavior and properties of the chip core to meet the specific requirements of your design.

It is important to note that the chip core may incorporate multiple macros, which are self-contained functional blocks within the core. These macros can be instantiated and interconnected to create a complex and cohesive chip design. Each macro may have its own specific configuration and characteristics, contributing to the overall functionality and capabilities of the chip core.

By appropriately configuring the environment variables and macros within the chip core, you can effectively tailor the behavior, features, and performance of the integrated circuit design in the OpenLane workflow.

5.1.3.1 Core Configuration File

The core of an integrated circuit design in the OpenLane workflow involves the utilization of various important environment variables. These variables play a crucial role in configuring and customizing the behavior of the core. Here are some of the main environment variables commonly used:

- **VERILOG_FILES**: It specifies the Verilog source files that are part of the core, The 'include directive is not supported.
- **VERILOG_FILES_BLACKBOX**: It is used to treat Verilog files as black boxes. It is necessary to add each Verilog file corresponding to the macro module used in the core to this list.
- **EXTRA_LEFS**: It specifies the LEF files associated with the pre-hardened macros incorporated into the core design.
- **EXTRA_LIBS**: This variable specifies the Library files of pre-hardened macros used in the current design. It aids in improving timing analysis and provides additional information for accurate characterization. This variable is optional.
- **EXTRA_GDS_FILES**: It is used for specifying GDSII files associated with pre-hardened macros integrated into the core.
- **SYNTH_READ_BLACKBOX_LIB**: This variable, specified true/false, is set to true when using any standard cells directly in the design. It indicates that the design does not purely function at the register transfer level.
- **MACRO_PLACEMENT_CFG**: If you wish to manually place the macros in specific locations, this variable provides the path to a file containing a line-break delimited list of instances and positions. The format for each line is instance_name X_pos Y_pos Orientation .

By appropriately setting these environment variables, you can control the inclusion of Verilog files, handle black-box modules, incorporate pre-hardened macros, define library and technology files, and manage macro placement within the core. These variables offer flexibility and customization options for designing and configuring the core in the OpenLane workflow.

The key variables employed to address PDN and LVS concerns in the design are FP_PDN_CORE_RING , VDD_NETS , GND_NETS , FP_PDN_MACRO , _HOOKS , SYNTH_USE_PG_PINS_DEFINES , FP_PDN_HPITCH , FP_PDN _VPITCH , FP_PDN_HOFFSET , and FP_PDN_VOFFSET .

The ALU core configuration file is shown below.

5.1 Macro-Cells

```json
{
"DESIGN_NAME": "ALU_8Bit",
"VERILOG_FILES": "dir::src/*.v",
"CLOCK_PORT": null,
"DESIGN_IS_CORE": true,

"MACRO_PLACEMENT_CFG": "dir::macro.cfg",

"VERILOG_FILES_BLACKBOX":["dir::V_BB/Adder8Bit.v",
"dir::V_BB/LeftShifter8Bit.v",
"dir::V_BB/RightShifter8Bit.v",
"dir::V_BB/Subtractor8Bit.v"],


"EXTRA_LEFS":["dir::LEF/Adder8Bit.lef",
"dir::LEF/LeftShifter8Bit.lef",
"dir::LEF/RightShifter8Bit.lef",
"dir::LEF/Subtractor8Bit.lef"],

"EXTRA_GDS_FILES":["dir::GDS/Adder8Bit.gds",
"dir::GDS/LeftShifter8Bit.gds",
"dir::GDS/RightShifter8Bit.gds",
"dir::GDS/Subtractor8Bit.gds"],

"FP_PDN_CORE_RING":true,
"FP_SIZING": "absolute",
"DIE_AREA": "5 0 115 210",
"FP_CORE_UTIL":80,

"FP_PDN_CORE_RING":true,
"VDD_NETS": "VPWR",
"GND_NETS": "VGND",
"FP_PDN_MACRO_HOOKS":["Adder8Bit VPWR VGND VPWR VGND,
    LeftShifter8Bit VPWR VGND VPWR VGND, RightShifter8Bit VPWR
    VGND VPWR VGND, Subtractor8Bit VPWR VGND VPWR VGND"],

"SYNTH_USE_PG_PINS_DEFINES": "USE_POWER_PINS",
"FP_PDN_HPITCH":20,
"FP_PDN_VPITCH":40,
"FP_PDN_HOFFSET":0,
"FP_PDN_VOFFSET":10
}
```

These variables play a crucial role in defining the power and ground nets, incorporating macro hooks for PDN integration, enabling the use of power and ground pin definitions in synthesis, and specifying the pitch and offset values for the PDN grid. By carefully configuring these variables, designers can effectively manage power distribution and ensure proper electrical connectivity, ultimately contributing to the overall performance and reliability of the design.

- **P_PDN_MACRO_HOOKS**: This variable is used to specify the PDN integration.It contain additional information or customizations required for specific macros in the design, the structure of the variable follows the format:
  <instance_name> <vdd_net> <gnd_net> <vdd_pin> <gnd_pin>
- **FP_PDN_HPITCH**: It defines the horizontal pitch or spacing between adjacent power or ground tracks in the PDN. It determines the density and distribution of power and ground connections across the chip.
- **FP_PDN_VPITCH**: It defines the vertical pitch or spacing between adjacent power or ground tracks in the PDN. Similar to FP_PDN_HPITCH, it influences the density and distribution of power and ground connections in the chip.
- **FP_PDN_HOFFSET**: This variable specifies the horizontal offset or displacement of the power and ground tracks in the PDN. It allows fine-tuning of the position of power and ground connections to optimize the power distribution and mitigate potential issues.
- **FP_PDN_VOFFSET**: It represents the vertical offset or displacement of the power and ground tracks in the PDN. Like FP_PDN_HOFFSET, it enables precise adjustment of the position of power and ground connections for improved power integrity.
- **Core design considerations**: A Core module in OpenLane begins its creation just like any other module, by initiating a new project through the terminal.

As a good practice, it is recommended to have all the necessary files within the project. This includes organizing the project with folders such as the "src" folder for source files, the "GDS" folder for GDSII files, the "LEF" folder for LEF files, and the macro ".cfg" file. Figure 5.2 illustrates the organized directory.

After creating the necessary folders, you need to add the corresponding files to each folder. This includes adding the LEF, GDSII, and Verilog files for each macro module. In the case of the ALU, you should add the files for the Adder8Bit, Subtractor8Bit, LeftShifter8Bit, and RightShifter8Bit modules to their respective folders. The GDSII and LEF files of the hardened modules are located in the directory:
.../runs/<tag>/results/final/ as illustrated in Figure 4.2.

The configuration variables VERILOG_FILES_BLACKBOX , EXTRA _LEFS , and EXTRA_GDS _FILES , should have a structure similar to:

5.1 Macro-Cells

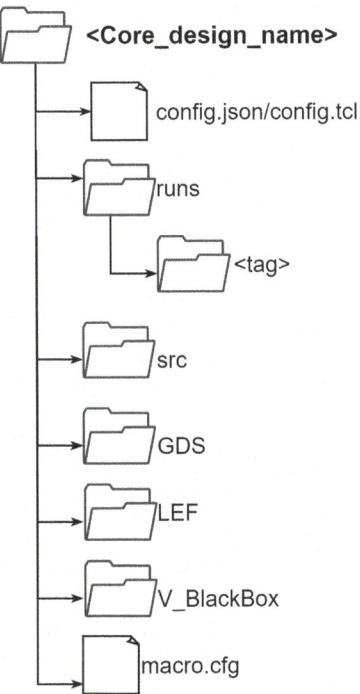

Fig. 5.2 The directory structure of a Core design with macros.

```
"VERILOG_FILES_BLACKBOX":
["dir::V_BB/Adder8Bit.v",
"dir::V_BB/LeftShifter8Bit.v",
"dir::V_BB/RightShifter8Bit.v",
"dir::V_BB/Subtractor8Bit.v"],
"EXTRA_LEFS":
["dir::LEF/Adder8Bit.lef",
"dir::LEF/LeftShifter8Bit.lef",
"dir::LEF/RightShifter8Bit.lef",
"dir::LEF/Subtractor8Bit.lef"],
"EXTRA_GDS_FILES":
["dir::GDS/Adder8Bit.gds",
"dir::GDS/LeftShifter8Bit.gds",
"dir::GDS/RightShifter8Bit.gds",
"dir::GDS/Subtractor8Bit.gds"],
```

The macro placement file is responsible for defining the precise coordinates at which each macro will be placed within the design. Specified by the MACRO _PLACEMENT_CFG configuration variable, this file consists of a line-delimited list of macro instances, their respective X and Y positions, and the desired orientation. The configuration variable looks as:

```
"MACRO_PLACEMENT_CFG": "dir::macro.cfg"
```

while the placement configuration file (macro.cfg) looks as:

```
Adder8Bit 50 20 N
LeftShifter8Bit 50 65 N
RightShifter8Bit 50 105 N
Subtractor8Bit 50 155 N
```

Please note that the macro placement file adheres to the instance_name X_pos Y_pos Orientation format. It specifies the name of each macro instance along with its corresponding X and Y positions and the desired orientation. This format ensures accurate control over the placement of macros within the design. By following this convention, designers can effectively organize and optimize the layout of macros, resulting in improved performance and connectivity in the integrated circuit.

Related to RTL modifications, it is necessary to add across each module (macros) that are instantiated in the top module (core) the comment ///sta-blackbox . e.g.

```
///sta-blackbox
module Adder8Bit(input [7:0] a, b, output [7:0] sum, `ifdef
    USE_POWER_PINS inout VPWR, inout VGND `endif);
    // 8-bit adder module
    assign sum = a + b;
endmodule
```

To ensure a proper PDN connection, it is essential to incorporate the power and ground pins in both the top module and its instantiations. This ensures that the necessary power and ground signals are properly routed throughout the design, facilitating a reliable and efficient power distribution.

Therefore, the resulting RTL code of the ALU, after incorporating the power and ground pins, can be found at [52]

The name of the voltage and ground pins must match at all times.

As a result of the steps mentioned above, an ALU is obtained which consists of 4 macros (Addition, Subtraction, shift left and shift right), Figure 5.3 shows the implemented circuit and clearly shows the 4 implemented logic blocks, we recommend to carefully review the restrictions mentioned above because they will be useful if you want to make an IC with macros in OpenLane.

5.2 SRAM Cells

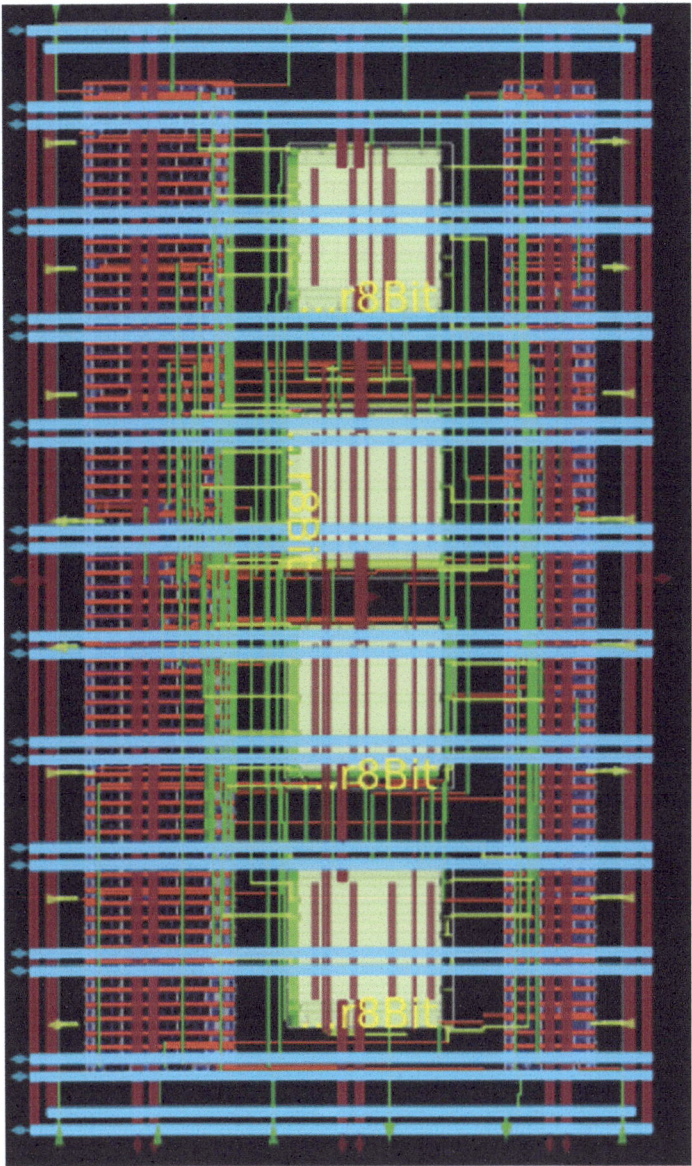

Fig. 5.3 Layout view of the designed ALU using OpenROAD GUI.

5.2 SRAM Cells

Memory plays a crucial role in digital systems, enabling quick and efficient data storage and retrieval. Among the various memory types, SRAM are widely used due to their high-speed access and compatibility with digitally integrated circuits. In this chapter, we delve into the integration of SRAM memories into digital systems.

SRAM memory arrays are organized in a grid-like structure, with rows and columns forming the foundation of the architecture. Each row represents a wordline, while each column corresponds to a bitline. This arrangement allows for efficient storage and retrieval of data. When a specific memory cell needs to be accessed, the corresponding wordline is activated to select the desired row, and the bitline connected to the target column is activated to read or write data. The intersection of a row and column represents an individual memory cell, typically consisting of a flip-flop circuit. The use of rows and columns in SRAM memory arrays enables simultaneous access to multiple cells, facilitating fast and parallel operations. Figure 5.4 illustrates the organization of the cells in an SRAM memory.

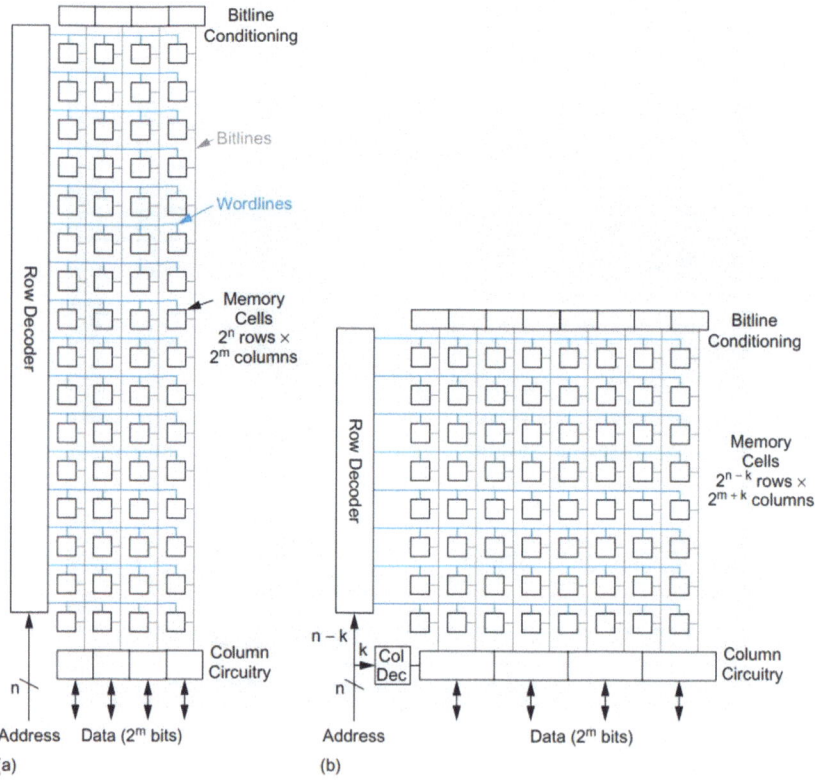

Fig. 5.4 Memory array architecture from [53].

The SRAM cell is the fundamental building block of RAM architecture. It is responsible for storing a single bit of data in a stable and reliable manner. The

5.2 SRAM Cells

most commonly used SRAM cell architecture is the 6-Transistor (6T) cell, which offers a balance between area efficiency and stability. The 6T SRAM cell consists of two inverters and two access transistors as Figure 5.5 shows. The inverters form a latch configuration, where the feedback loop of each inverter provides the necessary positive feedback to maintain the stored data. The access transistors act as switches that control the flow of data in and out of the cell.

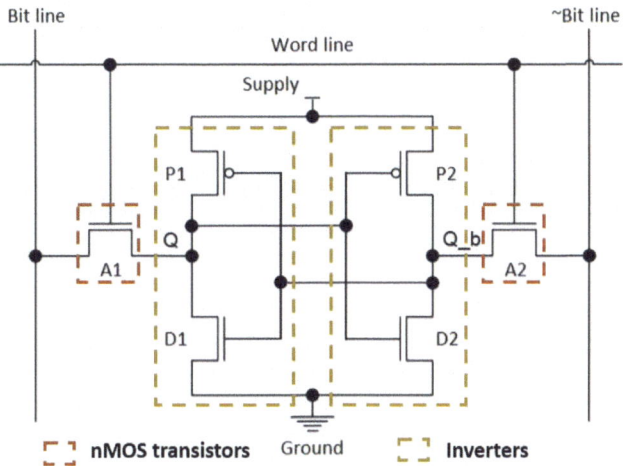

Fig. 5.5 The standard 6T SRAM cell.

To write data into the SRAM cell, the bitlines are precharged to a specific voltage level. The data to be stored is then driven onto the bitlines, and the access transistors of the targeted cell are activated, allowing the data to be latched into the cell. In the read operation, the access transistors are again activated, connecting the cell to the bitlines. The voltage levels on the bitlines are then compared to determine the stored data.

In the context of SRAM cell design, factors like leakage current, noise susceptibility, and process variations are carefully considered during the layout creation process. Unlike other digital systems that can be created using RTL to GDSII flow, SRAM memories require a manual layout design approach due to their unique requirements. Designers must pay close attention to transistor sizing, spacing, and routing to optimize stability and reliability.

To mitigate issues related to leakage current, proper transistor sizing is crucial. By carefully selecting the dimensions of transistors within the SRAM cell, designers can minimize leakage paths and reduce power consumption. Moreover, noise-reduction techniques, such as shielded routing and proper placement of power and ground lines, can help mitigate the impact of external noise sources and improve the stability of the SRAM cell.

Process variations, which can introduce variations in transistor characteristics, pose a challenge to SRAM cell design. Designers employ techniques such as redundancy, redundancy repair, and error correction codes to compensate for these variations and enhance the reliability of the memory. Additionally, advanced Design For Manufacturing (DFM) methodologies and statistical analysis are applied to account for process variations and improve yield.

Given the criticality of stability and reliability in SRAM cell design, the manual layout process allows designers to have fine-grained control over the placement and routing of components. This level of manual intervention enables the optimization of various design parameters to achieve the desired stability, reliability, and performance characteristics for the SRAM memory.

Other variations of SRAM cells, such as the 8-transistor (8T) and 10-transistor (10T) cells, offer improved stability and reduced susceptibility to certain failure modes. However, these cells generally come at the expense of increased area and complexity.

5.3 SKY130 SRAM Cell

The SKY130 PDK is a collection of files and libraries that provide the necessary information for designing integrated circuits using the SKY130 process technology. It includes design rules, device models, and technology files that enable circuit designers to develop and simulate their designs. The SKY130 PDK is specifically designed for the Google SkyWater 130nm process, which offers a low-cost, open-source alternative for custom chip manufacturing.

OpenRAM is an open-source memory compiler used for generating SRAM designs. It provides a flexible and configurable framework for creating different SRAM architectures based on specific requirements. OpenRAM allows for easy characterization and optimization of memory designs by automating the generation of layout, timing, and power parameters.

The SKY130 PDK, in conjunction with OpenRAM, has been used to generate three different SRAM configurations: 8×1024, 32×256, and 32×512. These SRAM designs were initially generated using OpenRAM's memory compiler and then manually fixed to address any DRC and other errors, ensuring they adhere to the SKY130 process technology requirements and avoid potential manufacturing issues. The SKY130 PDK includes a library with these pre-designed and characterized SRAM blocks, available in Verilog, LEF, and GDSII file formats, facilitating their seamless integration into system designs. Designers can instantiate these memory macros in their Verilog code and connect them to other components, efficiently incorporating SRAM functionality into their systems.

5.4 SRAM Macros with OpenLane, FIFO Memory Example

First-In, First-Out (FIFO) memory is a fundamental component in digital systems for managing data flow. It ensures that the data items stored are retrieved in the same order they were inserted. SKY130 SRAM macros, included in the SKY130 PDK, offer pre-designed SRAM blocks that serve as a foundation for implementing FIFO memory in integrated circuits. In this section, we will explore the concept of FIFO memory and its applications. Additionally, we will enhance the basic SRAM memory with management logic to transform it into a fully functional FIFO memory.

5.4.1 SKY130 SRAM Library

Creating a SRAM module is a complex task that requires expertise in VLSI design and a deep understanding of semiconductor processes. The intricate nature of SRAM design involves considerations such as device sizing, timing constraints, power optimization, noise immunity, and layout intricacies. Designing a robust and high-performance SRAM necessitates in-depth knowledge of circuit design principles, transistor-level design, digital logic, and memory architecture. However, in the context of the course utilizing SKY130 SRAM macros, students can leverage the hard work and expertise of experienced VLSI designers who have already crafted these macros. By utilizing these pre-designed macros, students can focus on understanding the functional aspects and management logic of SRAM modules, without delving into the intricacies of full SRAM design from scratch. This enables students to gain hands-on experience in VLSI and memory design while leveraging the SKY130 SRAM macros as a valuable learning resource.

The SKY130 PDK offers a collection of SKY130 SRAM macros (8×1024, 32×256, and 32×512). These macros are readily available building blocks that simplify the design process by providing pre-designed SRAM modules. To locate these SRAM macros, one must navigate to the PDK path, typically a hidden folder named ./volare in the home directory. Within this directory, you can find the SKY130A and SKY130B PDKs, with SKY130A being the current version. The SRAM files, including the GDS, LEF, and Verilog representations, can be found in the directory SKY130A where the PDK was installed.

To start utilizing the SRAM macros, the initial step is to create a new project in OpenLane, following the macro structure shown in Figure 5.2. Then, the GDS, LEF, and Verilog files of the desired SRAM macro need to be added to the project.

5.4.2 RTL SKY130 SRAM

To ensure smooth operation and prevent syntax errors, certain modifications need to be made to the original SRAM Verilog files. First, it is necessary to include the following lines at the top of the file:

```
`define USE_POWER_PINS
/// sta-blackbox
```

By adding these lines, we enable the utilization of power pins and indicate that the design is a black-box. Additionally, it is important to make a specific configuration change in a parameter to ensure proper functionality. The parameter NUM_WMASKS needs to be modified from 1 to 2. This adjustment will align the input of the GDSII SRAM file, referred to as wmask0[0], with the generated netlist input, which is called wmask0[0]. Failing to make this change can result in an LVS error during verification. OpenLane detects the discrepancy between the netlist input, which would be generated as wmask0 without the modification, and the expected input wmask0[0]. Therefore, modifying NUM_WMASKS ensures the correct matching of inputs and prevents potential errors during the design process.

As a final step, it is crucial to comment out the lines containing the #(T_HOLD) dout syntax within the Verilog file. By placing a comment symbol // at the beginning of these lines, we ensure that they are not interpreted as active code during the compilation and simulation processes. Commenting out these lines helps avoid any potential conflicts or syntax errors that may arise due to the specific syntax used in those lines. After making the necessary modifications, the Verilog file is visible and can be found at [54]

5.4.3 RTL FIFO Management Block

To create a FIFO memory based on an SRAM, it is crucial to understand the individual functionalities of both SRAM and FIFO. SRAM is a type of volatile memory that stores digital data in a static form, meaning it retains its contents as long as power is supplied. It consists of memory cells organized in a matrix-like structure, each cell is capable of storing a single bit of information. SRAM provides fast read and write access, making it suitable for cache memories and other high-performance applications.

On the other hand, a FIFO memory is a data structure that follows a strict queuing discipline, where the data that enters first is the first to be removed. It operates on the principle of enqueueing (writing) data at one end and dequeuing (reading) data from the other end. FIFOs are commonly used in scenarios where data needs to be buffered or transferred between systems with different data rates.

5.4 SRAM Macros with OpenLane, FIFO Memory Example

To create a FIFO memory based on SRAM, additional logic is added to the base SRAM memory. This logic manages the read and write operations, controls the enqueueing and dequeuing of data, and ensures the proper ordering of data elements. By combining the reliable and efficient storage capabilities of SRAM with the queuing discipline of a FIFO, we can design a memory structure that enables the systematic and orderly flow of data.

The utilization of SKY130 SRAM macros, which provide pre-designed and verified SRAM building blocks, simplifies the development of a FIFO memory. These macros offer a ready-to-use SRAM implementation, allowing designers to focus on integrating the necessary logic to enable FIFO functionality. By leveraging the SKY130 SRAM macros within the OpenLane environment, students can gain hands-on experience in designing and implementing FIFO memories efficiently and effectively.The code in [55] represents a complete implementation of a FIFO, including both the FIFO management logic and the SRAM block.

5.4.4 FIFO Configuration Files

The Figure 5.6 shows the GDSII file of the SRAM memory plus the management FIFO.

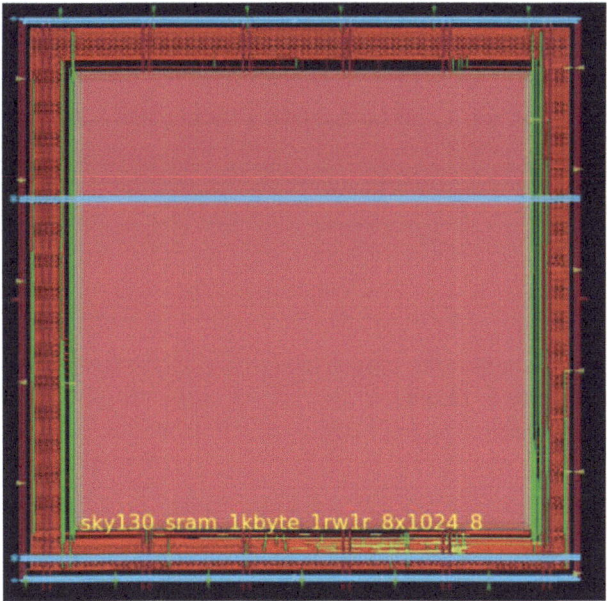

Fig. 5.6 FIFO memory.

The configuration file for generating the FIFO memory is:

```
{
"DESIGN_NAME": "FIFO",
"VERILOG_FILES": "dir::src/*.v",
"CLOCK_PORT": "clk",
"CLOCK_PERIOD": 10.0,
"DESIGN_IS_CORE": true,
"FP_PDN_CORE_RING": true,

 "MACRO_PLACEMENT_CFG": "dir::macro.cfg",
"VERILOG_FILES_BLACKBOX":["dir::verilog/sky130_sram_1kbyte
_1rw1r_8x1024_8.v"],
"EXTRA_LEFS":["dir::lef/sky130_sram_1kbyte_1rw1r_8x1024_8.
lef"],
"EXTRA_GDS_FILES":["dir::gds/sky130_sram_1kbyte_1rw1r_8x1024_8
.gds"],
"EXTRA_LIBS":
↪ ["dir::lib/sky130_sram_1kbyte_1rw1r_8x1024_8_TT_1p8V_25C
.lib"],

"FP_SIZING": "absolute",
"DIE_AREA": "0 0 550 550",
"FP_CORE_UTIL":80,

"VDD_NETS": "vccd1",
"GND_NETS": "vssd1",
"FP_PDN_MACRO_HOOKS":["sky130_sram_1kbyte_1rw1r_8x1024_8 vccd1
↪ vssd1 vccd1 vssd1"],
"SYNTH_USE_PG_PINS_DEFINES": "USE_POWER_PINS",
"FP_PDN_HPITCH":350,
"FP_PDN_VPITCH":100,
"FP_PDN_VSPACING":5,
"FP_PDN_HOFFSET":10,
"FP_PDN_VOFFSET":13,

"RUN_KLAYOUT_XOR": false,
"ROUTING_CORES":24

}
```

Chapter 6
Exploring OpenLane through Case Studies and Exercises

This chapter aims to provide practical examples and exercises that guide readers through the physical design flow using OpenLane. It covers a range of projects, from mathematical cores to communication controller modules and processors. Additionally, exercises are included to help optimize system performance, power consumption, or area.

This chapter explores five distinct projects: a Pseudo Random Generator, a Double Floating-Point Unit, an I2C Master/Slave module, AES-128 Encryption, and a basic RISC Processor. The RTL code for these designs was sourced from OpenCores.org, one of the largest and most recognized open-source hardware communities, with over 370,000 users contributing to a wide variety of digital designs.

Now, with the advent of OpenLane and open-source PDKs, the opportunity to implement complete systems on silicon has never been more accessible. This chapter aims to bridge the gap between design and physical implementation, empowering designers to take their ideas from RTL to silicon with open-source tools and technologies.

6.1 Pseudo Random Generator

This example presents the translation of a Non-Linear Pseudo-Random Generator (NLPRG) from RTL to GDSII. The NLPRG generates a pseudo-random sequence with a period of 2^n, where n represents the number of registers in the system. In this implementation, the layout uses a total of 8 registers to produce the sequence.

An NLPRG is a type of random number generator designed to produce a sequence of values that, while deterministic, appears random. Unlike Linear Feedback Shift Registers (LFSRs), which generate sequences based on linear functions, NLPRGs rely on non-linear feedback mechanisms, providing an added layer of unpredictability. This makes them more resistant to pattern detection and statistical attacks, which is especially valuable in cryptographic applications and secure communication systems.

By utilizing non-linear feedback, the NLPRG avoids the limitations of purely linear sequences, offering a longer period and better distribution of generated values. In this project, an 8-register NLPRG is implemented, capable of generating a pseudo-random sequence with a period of 2^8. The non-linear approach ensures that the sequence behaves in a way that is difficult to predict or reverse-engineer, making it ideal for applications where security and randomness are critical, the Verilog source code of this project can be found at [56].

6.1.1 Configuration File

To begin the OpenLane project, first, start the Docker session by running the command make mount inside of the OpenLane directory. Next, add your design with ./flow.tcl -design <design_name> -init_design_config -add_to_designs, replacing <design_name> with your desired design name. After this step, navigate to the src folder inside of the design folder and add all the Verilog files of the project, nlprg8.v, and dff.v in this case.

Next, in the configuration JSON file, replace the value of the DESIGN_NAME with the name of the top module "nlprg8" in this example. As in the previous example in Chapter 4, we also need to add the following constraints: FP_SIZING, DIE_AREA, FP_CORE_UTIL, PL_RANDOM_GLB_PLACEMENT, PL_TARGET_DENSITY, and PL_BASIC_PLACEMENT. Additionally, use AREA 3 for the synthesis strategy. The final configuration in the JSON file should look like this:

```
{
        "DESIGN_NAME": "nlprg8",
        "VERILOG_FILES": "dir::src/*.v",
        "CLOCK_PORT": "ck",
        "CLOCK_PERIOD": 10.0,
        "FP_PDN_MULTILAYER": true,
        "FP_PDN_CORE_RING":0,
        "FP_SIZING":"absolute",
        "SYNTH_STRATEGY":"AREA 3",
        "DIE_AREA":"0 0 100 100",
        "FP_CORE_UTIL":1,
        "PL_RANDOM_GLB_PLACEMENT":1,
        "PL_TARGET_DENSITY":0.8,
        "PL_BASIC_PLACEMENT":1
}
```

6.1 Pseudo Random Generator   71

Finally, execute the physical design flow by running the following command: ./flow.tcl -design <design_name> . If the process completes successfully, you should see the message "[SUCCESS]: Flow complete." at the end of the terminal output.

6.1.2 Suggested Experiments

1. **Exploring Synthesis Strategies:** Run all available synthesis strategies against your design using the command line option -synth_explore . A comprehensive report will be generated in the directory runs/<tag>/reports/synthesis/ The report, named "exploration_analysis", will provide details on the performance of each synthesis strategy.
2. **Minimizing Layout Area:** Reduce the layout area as much as possible by adjusting the DIE_AREA parameter. Explore different configurations to achieve the smallest possible design footprint.
3. **Maximizing System Frequency:** Increase the operating frequency of the system to its highest possible value by optimizing relevant design parameters and constraints.
4. **Optimizing Physical Design Flow:** Based on the synthesis strategies report, identify the configuration with the best gate count and use that strategy to proceed with the physical design flow, ensuring optimal resource utilization.
5. **Running the Flow with Custom Tag:** Execute the design flow using the following command line -tag <name> . Replace <name> with the desired name for your output folder. This will help organize different runs.
6. **Customizing Port Placement:** Modify the placement of input and output ports by using the FP_PIN_ORDER_CFG parameter. Set the input ports at the bottom of the module and the output ports at the top to improve routing efficiency.
7. **Generating Heat Maps:** Using the GUI, generate heat maps for the following metrics: Placement Density, Power Density, Routing Congestion, and Estimated Congestion. These visualizations will provide valuable insight into the design's physical layout, highlighting areas of potential improvement in terms of placement and routing efficiency.

6.1.3 Layout View

To view the resulting layout graphically, launch the GUI by executing the following command: python3 gui.py –viewer openroad <PATH> Here, <PATH> refers to the route directory, which can be found at the beginning of the physical design process and is indicated by the message "[INFO]: Run Directory:", the resultant layout can be seen in Figure 6.1.

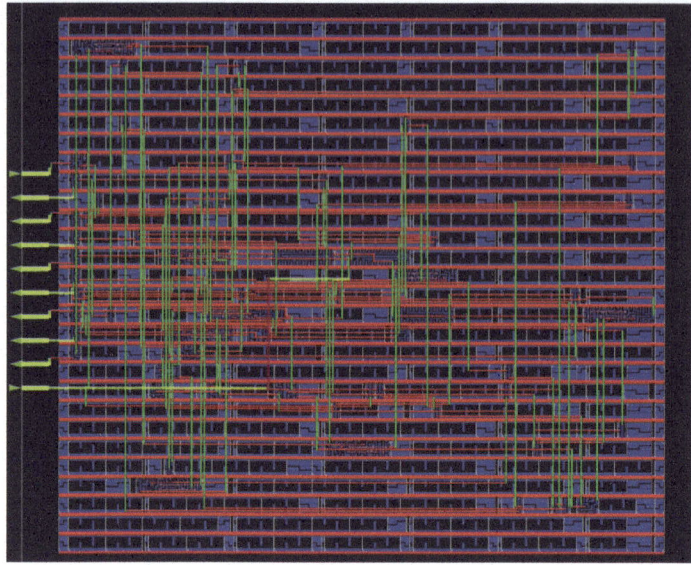

Fig. 6.1 Layout view of the designed Non-Linear Pseudo-Random Generator using OpenROAD GUI.

6.2 Double-Precision Floating Point Unit

The used double-precision floating-point unit (FPU) was designed to perform high-precision arithmetic operations following the IEEE-754 standard. The unit operates synchronously with a single global clock, updating all internal registers on the rising edge of the clock. A global reset signal is used to reset all registers, ensuring the unit can be initialized or reset easily.

The FPU supports four fundamental operations: addition, subtraction, multiplication, and division. Each operation has a specific clock cycle latency, with addition requiring 20 clock cycles, subtraction taking 21 clock cycles, multiplication using 24 clock cycles, and division requiring 71 clock cycles to complete. This design ensures precise handling of arithmetic tasks within the specified timing constraints.

This FPU supports all four rounding modes defined in the IEEE-754 standard: rounding to the nearest, rounding towards zero, rounding towards positive infinity ($+\infty$), and rounding towards negative infinity ($-\infty$). Additionally, it offers support for denormalized numbers, a feature that many floating-point units lack, as they often treat denormalized numbers as zero.

The unit is well-suited for high-performance computing applications where precision, compliance with IEEE standards, and efficient use of hardware resources are critical, the Verilog source code of this project can be found at [57].

6.2 Double-Precision Floating Point Unit

6.2.1 Configuration File

To begin the OpenLane project, first, start the Docker session by running the command make mount inside of the OpenLane directory. Next, add your design with ./flow.tcl -design <design_name> -init_design_config -add_to_designs, replacing <design_name> with your desired design name. After this step, navigate to the src folder inside of the design folder and add all the Verilog files of the project, fpu_double.v, fpu_add.v, fpu_div.v, fpu_mul.v, fpu_round.v, fpu_sub.v, and fpu_exceptions.v in this case.

Next, in the configuration JSON file, replace the value of the DESIGN_NAME with the name of the top module "fpu" in this example. Differing from the previous module, the FPU is a more complex design that involves hundreds of logic gates. Therefore, it is not necessary to add constraints such as FP_CORE_UTIL, PL_RANDOM_GLB_PLACEMENT, PL_TARGET_DENSITY, or PL_BASIC _PLACEMENT, as OpenLane provides a base configuration suited for medium-sized designs like this one. The only parameters modified from the default configuration were FP_SIZING and DIE_AREA. The final configuration in the JSON file should look like this:

```
{
    "DESIGN_NAME": "fpu",
    "VERILOG_FILES": "dir::src/*.v",
    "CLOCK_PORT": "clk",
    "CLOCK_PERIOD": 20.0,
    "FP_PDN_MULTILAYER": true,
    "FP_SIZING":"absolute",
    "DIE_AREA":"0 0 1500 1500"
}
```

Finally, execute the physical design flow by running the following command: ./flow.tcl -design <design_name> . If the process completes successfully, you should see the message "[SUCCESS]: Flow complete." at the end of the terminal output.

6.2.2 Suggested Experiments

1. **Minimizing Layout Area:** Optimize the layout by reducing the area as much as possible through adjustments to the DIE_AREA parameter. Experiment with different values to achieve the smallest possible design while maintaining functionality.
2. **Maximizing Frequency with Minimized Area:** After achieving the minimal DIE_AREA, aim to maximize the system's operating frequency. This experi-

ment focuses on pushing the frequency as high as possible while maintaining the previously optimized smaller area.
3. **Exploring Synthesis Strategies:** Use the -synth_explore command to run all available synthesis strategies on the input design. This will generate a comprehensive report, located in the runs/<tag>/reports/synthesis/ directory, under the name "exploration_analysis". Use this report to identify the optimal synthesis strategy for your design.
4. **Physical Design Flow for Best Frequency:** Based on the synthesis strategies report, execute the physical design flow using the configuration that yields the highest possible frequency for the system.
5. **Creating Individual FPU Operation Projects:** Create four separate OpenLane projects, one for each FPU operation: addition, subtraction, multiplication, and division. Follow the steps outlined in Chapter 5 on Macro Cells to implement and execute the physical design flow for each operation.
6. **Designing a Hierarchical FPU:** Build a hierarchical design for the entire FPU by following the hierarchical flow presented in Chapter 5 uses the files generated in the previous experiment. Ensure that you apply the necessary constraints: MACRO_PLACEMENT_CFG , VERILOG_FILES_BLACKBOX , EXTRA _LEFS , and EXTRA_GDS_FILES , to properly manage the macro placement and design hierarchy.

6.2.3 Layout View

To view the resulting layout graphically, launch the GUI by executing the following command: python3 gui.py –viewer openroad <PATH> Here, <PATH> refers to the route directory, which can be found at the beginning of the physical design process and is indicated by the message "[INFO]: Run Directory:", the resultant layout can be seen in Figure 6.2.

6.3 I2C Master/Slave

The I2C is a two-wire, bidirectional serial bus that offers a simple and efficient method for data exchange between devices. It is widely used in the consumer electronics and telecommunications sectors, as well as for board-level communication protocols. The used I2C Master Core provides an interface between a Wishbone Master and the I2C bus, making it an easy solution to integrate I2C functionality into any Wishbone-compatible system. For detailed specifications, the I2C protocol can be referenced on the Philips website.

The used core is fully compatible with the Philips I2C bus standard and supports multi-master operation, allowing multiple devices to communicate on the same bus. One of its key features is software programmable timing, enabling the user to adjust

6.3 I2C Master/Slave

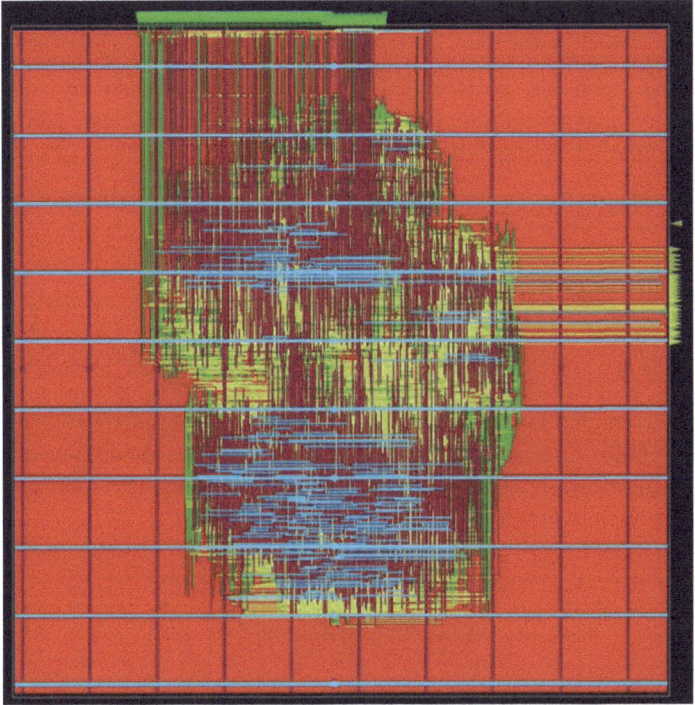

Fig. 6.2 Layout view of the designed Double-Precision Floating Point Unit using OpenROAD GUI.

communication speed as needed. The core also includes clock stretching and wait state generation, which ensures proper synchronization between devices by allowing slower devices to hold the clock signal until they are ready.

Data transfers are managed either by interrupts or by bit-polling, providing flexibility in data handling and processing. The core also supports arbitration loss detection, which triggers an interrupt and automatically cancels ongoing transfers, ensuring reliable communication in case of conflicts between multiple masters on the bus. The design facilitates the generation and detection of repeated start and stop conditions, as well as bus busy detection, which prevents conflicts during communication.

Supporting both 7-bit and 10-bit addressing, the core offers compatibility with a wide range of devices. It is fully static and synchronous, which allows for efficient integration into various digital systems. Additionally, the core is fully synthesizable, making it ready for implementation in hardware designs. This comprehensive feature set makes the I2C Master Core a versatile and robust solution for incorporating I2C communication in modern digital systems, the Verilog source code of this project can be found at [58].

6.3.1 Configuration File

To begin the OpenLane project, first, start the Docker session by running the command make mount inside of the OpenLane directory. Next, add your design with ./flow.tcl -design <design_name> -init_design_config -add_to_designs, replacing <design_name> with your desired design name. After this step, navigate to the src folder inside of the design folder and add all the Verilog files of the project, i2c_master_bit_ctrl.v, i2c_master_byte_ctrl.v, i2c_master_defines.v, i2c_master_top.v, and ms_core.v in this case.

Before making any changes to the configuration JSON file, it is important to modify the Verilog files of [58] by removing the timing constructor ("#"), as Yosys cannot synthesize it. Similar to the previous example, this design requires minimal adjustments to the configuration files. The key modifications are limited to the parameters CLOCK_PORT, DIE_AREA, and FP_SIZING. These changes ensure proper handling of the clock signal, area constraints, and floorplanning for the design. The final configuration in the JSON file should look like this:

```
{
    "DESIGN_NAME": "i2c_master_top",
    "VERILOG_FILES": "dir::src/*.v",
    "CLOCK_PORT": "wb_clk_i",
    "CLOCK_PERIOD": 15.0,
    "FP_PDN_MULTILAYER": true,
    "FP_SIZING":"absolute",
    "DIE_AREA":"0 0 600 600"
}
```

Finally, execute the physical design flow by running the following command: ./flow.tcl -design <design_name>. If the process completes successfully, you should see the message "[SUCCESS]: Flow complete." at the end of the terminal output.

6.3.2 Suggested Experiments

1. **Exploring Synthesis Strategies:** Use the command -synth_explore to run all available synthesis strategies on the input design. This will generate a comprehensive report, which can be found in the directory runs/<tag>/reports/synthesis/ under the name "exploration_analysis". Review this report to compare the different synthesis strategies and their impact on the design.
2. **Minimizing Area:** Select the synthesis strategy that minimizes area based on the generated report. Then, further reduce the area of the layout by adjusting the

DIE_AREA parameter. Experiment with different configurations to achieve the smallest possible layout while maintaining design functionality.
3. **Resolving Max Fanout Violations:** Address and eliminate any maximum fanout warning violations. This can be achieved by reviewing the fanout reports and optimizing the design to stay within the required fanout limits.
4. **Generating Heat Maps:** Using the GUI, generate heat maps for the following metrics: Placement Density, Power Density, Routing Congestion, and Estimated Congestion. These visualizations will provide valuable insight into the design's physical layout, highlighting areas of potential improvement in terms of placement and routing efficiency.

6.3.3 Layout View

To view the resulting layout graphically, launch the GUI by executing the following command: python3 gui.py –viewer openroad <PATH> Here, <PATH> refers to the route directory, which can be found at the beginning of the physical design process and is indicated by the message "[INFO]: Run Directory:", the resultant layout can be seen in Figure 6.3.

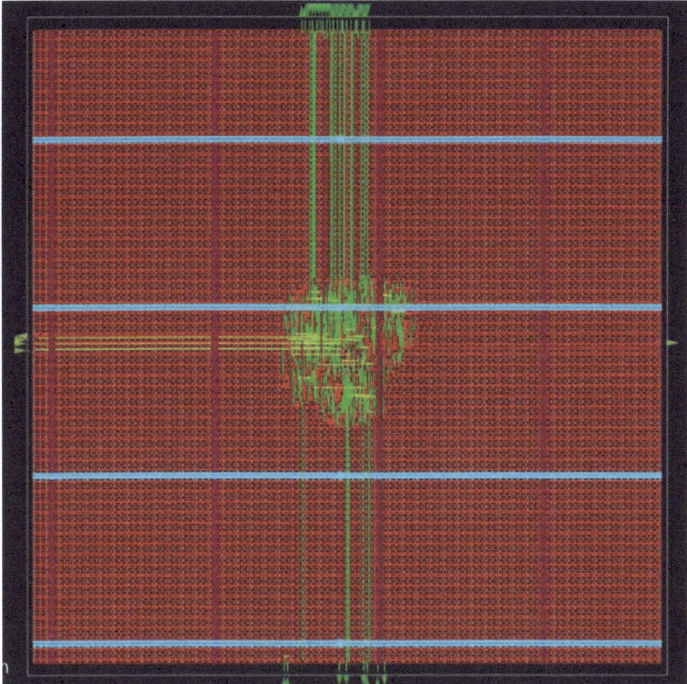

Fig. 6.3 Layout view of the designed I2C controller core using OpenROAD GUI.

6.4 AES-128 Encryption

The Advanced Encryption Standard (AES) is a widely used symmetric encryption algorithm that provides robust data protection through block ciphering. It operates on fixed-size blocks of data, offering a secure and efficient method of encrypting and decrypting information. This project provides two different implementations of the AES encryption core: a 128-bit and a 192-bit variant, both optimized for very low area applications, making them highly suitable for resource-constrained environments such as embedded systems, secure communications, and data storage.

The 128-bit AES core focuses on minimal area consumption and takes approximately 500 cycles to encrypt or decrypt a block of data. Similarly, the 192-bit AES core also prioritizes low area usage and achieves encryption or decryption in about 280 cycles. Both implementations avoid using memories to store the S-box and incorporate various architectural improvements to further reduce area consumption, making them highly efficient for applications where hardware resources are limited.

Unlike many AES cores, these implementations combine both the encryption and decryption functions in the same block, providing a versatile solution in a compact design. The cores are fully compliant with the AES standard as defined by the National Institute of Standards and Technology (NIST) and follow the required number of rounds for encryption—10 rounds for the 128-bit variant and 12 rounds for the 192-bit variant. Each round consists of byte substitution, row shifting, column mixing, and key addition.

The cores support both the 128-bit and 192-bit key lengths and offer flexible operation modes, allowing data to be processed either in Electronic Codebook (ECB) or Cipher Block Chaining (CBC) modes. Additionally, they support padding for data sizes that are not exact multiples of the block size, ensuring compatibility with a wide range of data streams.

These AES cores were developed using SystemC RTL and verified through Transaction Level Modeling (TLM). The design is fully synchronous, and Verilog synthesizable code is also provided, making these cores suitable for implementation on both FPGA and ASIC platforms. The low-area AES encryption cores are a highly efficient solution for secure, high-speed data encryption in constrained environments. The Verilog source code for these AES-128 and AES-192 low-area implementations can be found at [59].

6.4.1 Configuration File

To begin the OpenLane project, first, start the Docker session by running the command make mount inside of the OpenLane directory. Next, add your design with ./flow.tcl -design <design_name> -init_design_config -add_to_designs , replacing <design_name> with your desired design name. After this step, navigate to the src folder inside of the design folder and add all the Verilog files of the project,

6.4 AES-128 Encryption

aes.v, byte_mixcolum.v ,keysched.v ,mixcolum.v ,sbox.v ,subbytes.v ,timescale.v ,wb_aescontroller.v ,word_mixcolum.v in this case.

Next, in the configuration JSON file, replace the value of the DESIGN_NAME with the name of the top module "aes" in this example. The AES is a more complex design that involves hundreds of logic gates it is not necessary to add constraints such as FP_CORE_UTIL , PL_RANDOM_GLB_PLACEMENT , PL_TARGET_DENSITY , or PL_BASIC _PLACEMENT , as OpenLane provides a base configuration suited for medium-sized designs like this one. The only parameters modified from the default configuration were FP_SIZING and DIE_AREA . The final configuration in the JSON file should look like this:

```
{   "DESIGN_NAME": "aes",
    "VERILOG_FILES": "dir::src/*.v",
    "CLOCK_PORT": "clk",
    "CLOCK_PERIOD": 15.0,
    "FP_PDN_MULTILAYER": true,
    "FP_SIZING":"absolute",
    "DIE_AREA":"0 0 500 500" }
```

Finally, execute the physical design flow by running the following command: ./flow.tcl -design <design_name> . If the process completes successfully, you should see the message "[SUCCESS]: Flow complete." at the end of the terminal output.

6.4.2 Suggested Experiments

1. **Eliminating Max Fanout and Max Slew Violations:** Address and eliminate any max fanout warnings and max slew violations. This can be achieved by reviewing the synthesis reports and optimizing the design to stay within the required fanout and slew limits. Fine-tuning the design will ensure optimal signal integrity and timing performance.
2. **Exploring Synthesis Strategies:** Use the command -synth_explore to run all available synthesis strategies on the input design. A detailed report will be generated and can be found in the directory runs/<tag>/reports/synthesis/ under the name "exploration_analysis". Analyze this report to compare the impact of different synthesis strategies on your design.
3. **Minimizing Layout Area:** After selecting the synthesis strategy that consumes the least area based on the report, proceed to reduce the layout area further by configuring the DIE_AREA parameter. Experiment with various configurations to achieve the smallest area possible while maintaining functionality.
4. **Generating Heat Maps:** Leverage the GUI to generate heat maps for the following key metrics: Placement Density, Power Density, Routing Congestion, and Estimated Congestion. These visual maps will offer crucial insights into the de-

sign's layout, helping to identify potential improvements in terms of placement and routing efficiency.
5. **Executing the Physical Design Flow for AES-192:** Create a new OpenLane project for the AES-192 core. Execute the complete physical design flow by following the steps outlined in previous chapters. This will ensure that the AES-192 implementation is fully optimized and ready for synthesis and place-and-route stages.

6.4.3 Layout View

To view the resulting layout graphically, launch the GUI by executing the following command: python3 gui.py –viewer openroad <PATH> Here, <PATH> refers to the route directory, which can be found at the beginning of the physical design process and is indicated by the message "[INFO]: Run Directory:", the resultant layout can be seen in Figure 6.4.

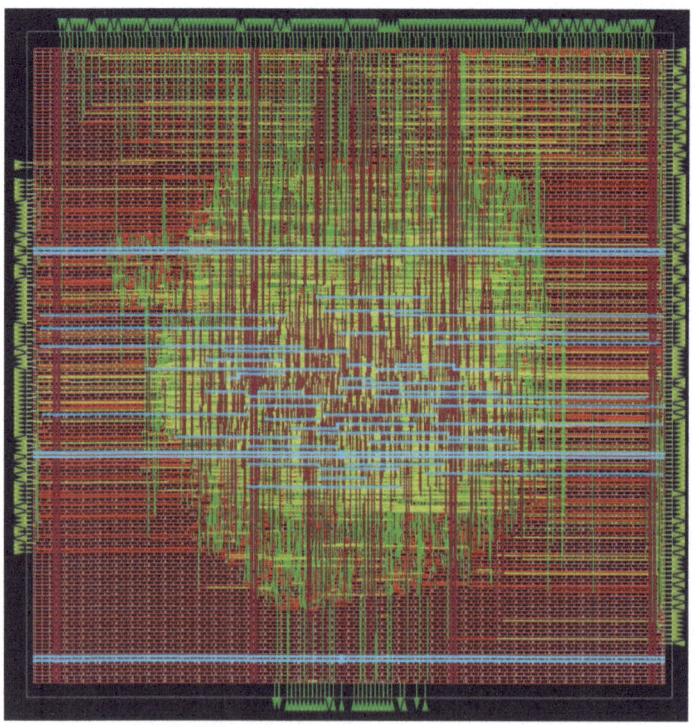

Fig. 6.4 Layout view of the implemented AES-128 using OpenROAD GUI.

6.5 RISC-V Single Cycle

The RISC-V RV32I Single Cycle processor is a compact yet powerful implementation of the RISC-V Instruction Set Architecture (ISA), specifically designed for applications that require a balance between performance and simplicity. This design is well-suited for educational purposes, embedded systems, and low-power applications. The processor operates in a single cycle, meaning that each instruction is completed in one clock cycle, providing a straightforward execution model while maintaining efficiency.

At the core of the processor is the ALU (Arithmetic Logic Unit), which performs the arithmetic and logic operations required by the instruction set. Supporting the ALU is the Register File, a set of 32 registers that stores the operands for the instructions and holds the results of operations. These registers are accessed using two read ports and one write port, allowing simultaneous data access and ensuring the smooth execution of instructions.

The processor's Memory Address Register interfaces with the system's memory to fetch instructions and store or load data as needed. The Immediate Generation unit is responsible for producing immediate values required by certain instructions, particularly those involving memory accesses and branching operations.

The Control Unit orchestrates the entire operation of the processor by decoding the instructions and generating control signals to ensure that each component functions correctly. The Type Decode and Control Decode units work together to interpret the instruction format and determine the appropriate actions for each instruction, whether it involves computation, memory access, or branching.

Memory operations are handled through an integrated RAM component, allowing for both data storage and retrieval as required by load/store instructions. The Branch Circuit plays a crucial role in handling conditional and unconditional jumps, ensuring that the program flow is correctly managed by updating the Program Counter (PC) when a branch is taken.

The Program Counter itself is a critical component, as it keeps track of the next instruction to be executed. After each cycle, the PC is updated to point to the next instruction in memory, unless a branch or jump instruction modifies its value.

This RISC-V processor is fully synthesizable with commercial tools and designed for efficient hardware implementation. Its modular structure allows for easy integration into larger systems, making it an ideal candidate for use in custom processors or as a foundational building block for more complex designs. The Verilog source code for this single-cycle RV32I implementation can be found at [60].

6.5.1 Configuration File

To begin the OpenLane project, start the Docker session by running the command make mount inside of the OpenLane directory. Next, add your design using the command ./flow.tcl -design <design_name> -init_design_config -add_to_designs , replacing <design_name> with your desired design name. After this step, navigate to the src folder inside of the design folder and add all the Verilog files of the project, ALU.v, PC.v, adder.v, arbiter.v, branch.v, byte_access.v, control.v, data_mem.v, immediate.v, mux1_2.v, mux2_4.v, ram.v, reg_file.v, top.v, unit.v, wrapper.v in this case.

Before modifying the configuration JSON file, some adjustments must be made to the Verilog files provided in [60] due to certain errors generated by Yosys that are not encountered in commercial tools. Specifically, lines 129 to 135 in the top.v file should be commented out, as well as line 9 in the branch.v file. As with previous examples, this design requires minimal changes to the configuration files. The key modifications involve updating the parameters CLOCK_PORT , DIE_AREA , and FP_SIZING . These adjustments ensure the proper configuration of the clock signal, area constraints, and floorplanning for the design. The final configuration in the JSON file should resemble the following:

```
{
    "DESIGN_NAME": "top",
    "VERILOG_FILES": "dir::src/*.v",
    "CLOCK_PORT": "clk",
    "CLOCK_PERIOD": 15.0,
    "FP_PDN_MULTILAYER": true,
     "FP_SIZING":"absolute",
    "DIE_AREA":"0 0 800 800"
}
```

Finally, execute the physical design flow by running the following command: ./flow.tcl -design <design_name> . If the process completes successfully, you should see the message "[SUCCESS]: Flow complete." at the end of the terminal output.

6.5.2 Suggested Experiments

1. **Creating Individual Projects:** Create four separate OpenLane projects, one for each main RISC-V module: adder, PC, control, unit, reg_file, immediate,ALU and branch. Follow the steps outlined in Chapter 5 on Macro Cells to implement and execute the physical design flow for each operation, execute the Synthesis Strategies (Use the -synth_explore command) execute the physical design flow using the configuration that yields the highest possible frequency per system.
2. **RAM memory substitution:** Modify the top module to incorporate a wrapper module, which encapsulates the original design. In this experiment, replace the RAM component within the design with the SkyWater RAM IP. This substitution will allow the design to utilize the SkyWater open-source memory block, ensuring better integration with the SkyWater process node and enhancing compatibility with open-source PDKs. Update the necessary connections in the wrapper module

6.5 RISC-V Single Cycle

to properly interface the SkyWater RAM IP, and verify the functionality of the overall system after integration. .

6.5.3 Layout View

To view the resulting layout graphically, launch the GUI by executing the following command: python3 gui.py –viewer openroad <PATH> Here, <PATH> refers to the route directory, which can be found at the beginning of the physical design process and is indicated by the message "[INFO]: Run Directory:", the resultant layout can be seen in Figure 6.5.

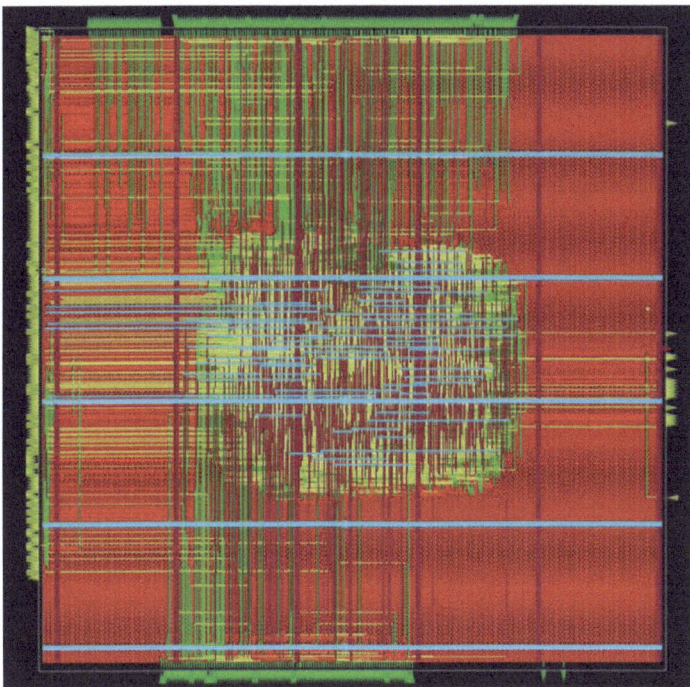

Fig. 6.5 Layout view of the designed RISC-V RV32I Single Cycle processor using OpenROAD GUI.

Chapter 7
Caravel

This chapter delves into the specifics of the Caravel open-source ASIC design framework. The initial section introduces Caravel, detailing its architecture and unique features, and guides the reader through the installation process. This part of the chapter provides a comprehensive understanding of Caravel's capabilities and how it fits into the broader landscape of open-source ASIC design tools.

The subsequent section focuses on Efabless, the platform that hosts Caravel. It discusses the Chipignite Program, the Multi-Project Wafer sponsored by Google, and Efabless' collaboration with GlobalFoundries. This section also elucidates the roles of SkyWater and Efabless in the open-source ASIC design ecosystem, providing readers with a broader context of the industry.

The latter part of the chapter provides a practical guide on using Caravel, covering topics such as project directory structure, repository integration, Verilog integration, and GPIO configuration. It also walks the reader through the process of layout integration and the RTL to GDSII flow with Caravel. The chapter concludes with a detailed discussion on SoC integration, including user area pins, logic analyzer pins, and GPIO pins, and provides an example of SoC integration. This blend of theoretical knowledge and practical instructions makes this chapter a valuable resource for both novices and experienced individuals in the field of ASIC design.

7.1 Introduction to Caravel

Caravel is a powerful and versatile open-source SoC design framework that has gained significant popularity in the field of VLSI design. It provides a complete and robust platform for designing and prototyping custom digital chips, enabling designers to create their own complex integrated circuits with ease.

At its core, Caravel consists of a fully functional open-source SoC design named "Caravel SoC" that serves as a foundation for creating custom chips. The Caravel SoC is built using the SkyWater Open Source PDK called SKY130, which provides a comprehensive set of technology files, device models, and design rules required for designing chips.

The development of Caravel began as an initiative to create an open-source platform that would facilitate the rapid design and verification of custom chips. It was initially conceived by Tim Edwards and other contributors at efabless corporation, a company focused on providing a platform for the design and production of semiconductor chips. Recently efabless, Google. and SkyWater foster open-source semiconductor design, the project aimed to democratize chip design by providing a community-driven, collaborative framework that would lower the barrier to entry for chip development.

The first version of Caravel was released in 2018, and since then, it has undergone significant enhancements and refinements through the active involvement of a vibrant community of designers, engineers, and enthusiasts. The collaborative nature of the Caravel project has fostered continuous innovation, resulting in a highly versatile and stable SoC platform that can be customized and adapted to a wide range of applications.

Caravel offers several compelling advantages that make it an attractive choice for designers venturing into custom chip development:

- **Open-source nature**: Caravel is built on the principles of open-source development, allowing users to freely access, modify, and redistribute the framework. This fosters collaboration, knowledge sharing, and a sense of community, enabling designers to benefit from collective expertise and contribute to the growth of the platform.
- **Comprehensive SoC design**: The Caravel SoC provides a feature-rich and complete system design that incorporates essential components, such as a microcontroller subsystem, memory, communication interfaces, and configurable I/Os. This eliminates the need for designers to start from scratch and enables them to focus on specific customizations and additions required for their applications.
- **Integration with SKY130 PDK**: Caravel seamlessly integrates with the SKY130 PDK, a widely used open-source process design kit. Leveraging this integration, designers can take advantage of the advanced process technologies, transistor models, and design rules provided by the SKY130 PDK to achieve optimal performance and reliability in their chip designs.
- **Pre-verified IP blocks**: Caravel includes a library of pre-verified intellectual property (IP) blocks that can be readily integrated into designs. These IP blocks range from standard interfaces to specialized components, enabling designers to accelerate their development process by reusing validated and proven functionality.

In the subsequent chapters, we will delve deeper into the architecture, design flow, and customization options offered by Caravel, empowering you to harness the full potential of this remarkable open-source SoC platform.

7.1.1 Caravel Architecture

The Efabless Caravel chip is a ready-to-use test harness for creating designs with the Google/Skywater 130nm Open PDK. The Caravel harness comprises base functions supporting IO, power, and configuration as well as drop-in modules for a management SoC core, and an approximately 2.92mm × 3.52mm open project area, called user project area, for the placement of user IP blocks [61].

Caravel is a template SoC for Efabless Open MPW and chipIgnite shuttles based on the Sky130 node from SkyWater Technologies. The current architecture of the SoC is illustrated in Figure 7.1.

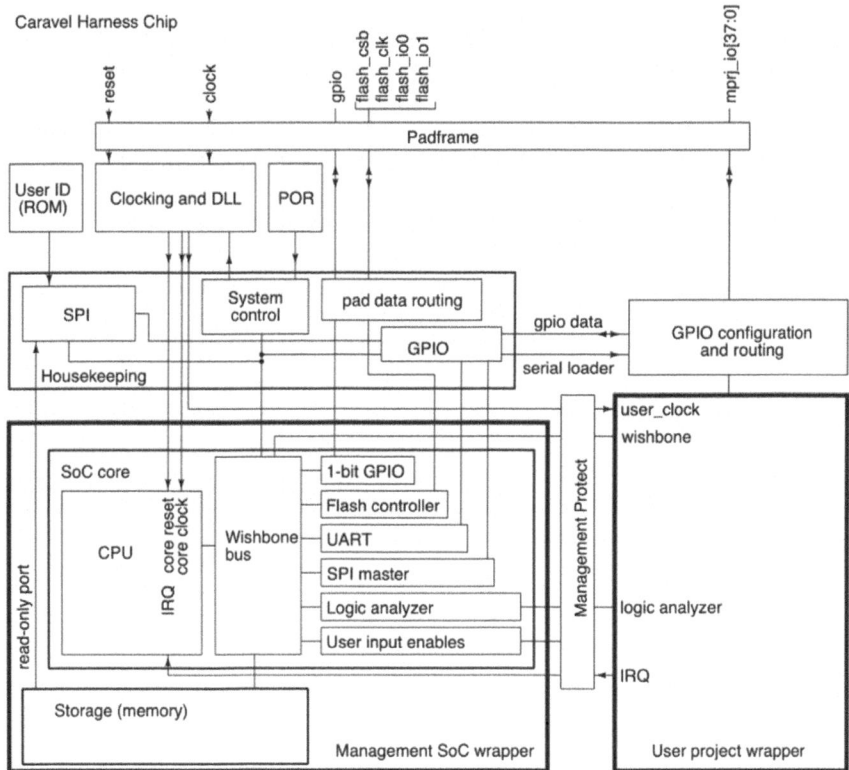

Fig. 7.1 Caravel SoC architecture [61].

Caravel consists of a harness frame and two wrappers for drop-in modules, designated for the management area and the user project area. The harness includes essential components such as the clocking module, DLL, user ID, housekeeping SPI, Power-On Reset, and GPIO control.

7.1.1.1 Management Area

In the context of Caravel, the management area functions as a pre-verified, hardened section of the chip. It assists in managing the user project area and provides debug capabilities. It has a standardized interface, so users can focus on the design and verification of their own projects without having to worry about the overall system management.

The management area in Caravel is a SoC generated using LiteX, a flexible and modular Python library for digital design. This SoC includes various components:

- **VexRiscv core**: This is the heart of the SoC. VexRiscv is a highly configurable and modular RISC-V processor implementation. It acts as the main computational unit.
- **Memory**: This is the storage section of the SoC, which could include both RAM and ROM. It is used to store data and instructions for the VexRiscv core.
- **Flash Controller**: This component is responsible for managing the flash memory of the SoC. Flash memory is a type of non-volatile memory, meaning it retains data even when power is switched off. The controller allows the SoC to read from and write to the flash memory.
- **Serial Peripherals**: These are devices that communicate with the VexRiscv core and other components of the SoC using serial communication, a form of data transmission where data is sent one bit at a time. Examples of serial peripherals could include UARTs, SPI devices, and I2C devices.

7.1.1.2 User Project Area

This is the user space. It has a limited silicon area 2.92mm × 3.52mm as well as a fixed number of I/O pads 38 and 4 power pads. The user space has access to the following utilities provided by the management SoC:

- 38 IO Ports
- 128 Logic analyzer probes
- Wishbone port connection to the management SoC wishbone bus.

7.1.2 Features

The main features of caravle are listed below, Figure 7.2 shows both the mnagment area and the user area.

- VexRiscv core with debug port
- 2 kB SRAM plus 1 kB of DFFRAM
- XIP SPI Flash controller
- UART, SPI and GPIO ports
- 128 port logic analyzer

7.1 Introduction to Caravel

- Counter / timer
- 32-bit Wishbone bus extending to the user project area
- 6 user interrupts
- $10mm^2$ Silicon

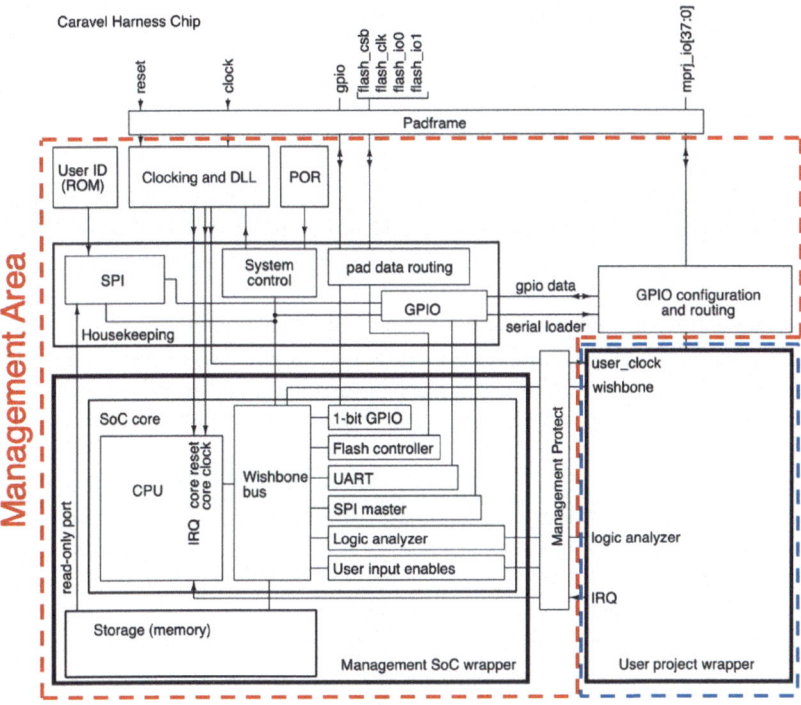

Fig. 7.2 Management Area and User Area of Caravel [61].

7.1.3 Caravel Installation

To begin your project in the OpenMPW program sponsored by Google, the first step is to create a new repository on GitHub, based on the caravel template at [62]. This template provides a starting point for your project, including essential files and directories specific to the Caravel design flow.

When creating the repository, ensure that it is set to public visibility, allowing others to view and collaborate on your project. Additionally, including a README file in your repository is highly recommended. The README serves as a documentation hub where you can provide an overview of your project, explain its purpose, outline the design goals, and offer instructions on how to use and contribute to the project.

By using the caravel_user_project template and maintaining a public repository with a comprehensive README, you establish a solid foundation for collaboration, knowledge sharing, and engagement within the MPW program community. This enables others to understand your project, contribute their expertise, and foster a vibrant ecosystem of open-source silicon development.

After following the installation steps, which can be found in [63], your environment will be set up with essential components that are crucial for your project. The following key elements will be installed:

- **Caravel Lite**: This is a lightweight version of Caravel, it provides a streamlined version of the framework, allowing for efficient chip design and development. It offers a foundation for integrating various IP blocks, running simulations, and preparing the design for fabrication.
- **Management Core for Simulation**: This component encompasses the management core, which plays a vital role in facilitating simulations of your design. The management core provides the necessary infrastructure and interfaces to execute simulations effectively, enabling rigorous testing and verification of your chip's functionality.
- **OpenLane**: OpenLane is a powerful open-source digital design flow that aids in the hardening of your design. It automates several steps involved in the design process, including synthesis, placement, routing, and optimization. OpenLane ensures that your design meets the required specifications, constraints, and performance targets, making it ready for fabrication.
- **Process Design Kit**: The installation includes the Process Design Kit, which contains a collection of files, libraries, and design rules specific to the targeted process technology. The PDK provides essential information and guidelines necessary for designing chips compatible with the chosen fabrication process. It enables accurate layout design and ensures compliance with the process requirements.

To install the full version of Caravel instead of the default Caravel Lite, you can execute the following command before proceeding with the setup $ export CARAVEL_LITE=0 .

7.2 Efabless

Efabless is a crowdsourcing platform and marketplace for chip design. It corporation is an open innovation, hardware creation platform for "smart" products. Their community delivers the customized integrated electronics required for semiconductor and hardware system innovators to turn their product visions into a marketable reality.

7.2 Efabless

Today, Efabless focuses on on-demand and custom IP core . Efabless gives chip companies two ways to obtain analog and mixed-signal IP: it's possible to search for existing IP in their growing library of verified designs, but more importantly you can also request the Efabless community to design new IP and derivatives [64].

7.2.1 Chipignite Program

The chipIgnite program, develop by Efabless and a collaboration with SkyWater Technology, expands upon the SKY130-based open source chip manufacturing program sponsored by Google and supports private commercial designs that include non-open source IP. This initiative represents another step forward in the industry to broaden access to chip design by giving people the ability to more easily create and fabricate chips.

It offers to the designer which includes not only low-cost manufacturing, but also a development board and firmware stack to simplify design validation and test. The program also includes an optional automated open source design flow for implementing projects that enable users to generate layouts for their digital projects from RTL. In addition, the program provides users with $10mm^2$ of total project area with fabrication for projects using SkyWater's open source PDK.

The program is a good fit for users who want to create an initial prototype or proof-of-concept for an IP block or full SoC. The starting price of $9750 per project includes 100 QFN or 300 WCSP packaged parts and five evaluation boards. The chipIgnite shuttles also support users who are distributing initial boards or launching a pilot for their product. An option for 1000 WCSP parts at $20 each is available that enables the service to be used for early product builds [65].

7.2.2 Multi-Project Wafer Sponsored by Google

Google has been actively involved in promoting open-source silicon projects through its Open Multi Project Wafer (OpenMPW) program. The OpenMPW program, launched in partnership with silicon fab SkyWater and design platform Efabless, allows designers to submit their designs for fabrication into physical chips at zero cost.

The OpenMPW program has seen remarkable success since its inception, with hundreds of successful designs being produced. The program aims to foster collaboration, innovation, and accessibility in the semiconductor industry by providing a platform for designers to bring their open-source silicon projects to life.

7.2.3 Collaboration with GlobalFoundries

In a significant development, Google extended its OpenMPW program to include GlobalFoundries' 180 nm process node. The collaboration allows free and open-source silicon projects to leverage the GlobalFoundries PDK and send their designs for production into silicon chips at no cost.

The GlobalFoundries PDK, based on the 180nm process node, provides designers with the opportunity to design for production without licensing fees. This collaboration further strengthens Google's commitment to supporting the open-source ecosystem and empowering designers to realize their innovative ideas.

7.2.4 The Role of SkyWater and Efabless

SkyWater, a trusted technology realization partner, and Efabless, a crowdsourcing design platform for custom silicon, play crucial roles in the implementation of the OpenMPW program. SkyWater manufactures the MPW shuttles, ensuring the fabrication of the submitted designs, while Efabless manages the submission process and provides the necessary design tools and resources.

Through Efabless's Caravel platform, a carrier RISC-V based System on Chip, designers can access a standardized test harness that supports their designs and reduces barriers to design, prototyping, and verification. Each project owner participating in the OpenMPW program receives four development boards with their design, along with additional parts for other uses.

7.3 Caravel User Project Directories

During this section, participants will explore the main directories within the Caravel project. These directories play crucial roles in the design and development process of Caravel.

- **gds directory**: The gds directory serves as a repository for the GDS files generated during each flow of OpenLane. These GDS files play a critical role as they represent the physical layout of the integrated circuits in a format that is suitable for fabrication.
- **lef directory**: Participants can find the LEF files. LEF files describe the detailed information about the library cells, such as their physical dimensions, pin locations, and routing information.
- **lib directory**: Stores the library files, which provide the necessary logic and analog primitives for designing integrated circuits.

- **openlane directory**: Serves as the equivalent of the "OpenLane/Design" directory. This directory is the designated area where users are required to create their specific OpenLane projects.
- **verilog/dv directory**: Contains the Verilog files related to design verification (DV). DV involves creating testbenches, test vectors, and test cases to verify the correctness and functionality of the design.
- **verilog/gl directory**: Dedicated to storing the gate-level Verilog files. These files represent the circuit design at the gate-level abstraction, where the logic functions are implemented using individual gates. The gate-level Verilog files provide a detailed representation of the circuit's structure and connectivity, facilitating further analysis, optimization, and verification processes during the design flow.
- **verilog/rtl directory**: Houses the RTL Verilog files. These files describe the behavior and functionality of the individual modules within the Caravel design.

Understanding the purposes and contents of these directories is essential for participants to effectively work with the Caravel project, enabling them to develop and customize Caravel-based designs with ease.

7.4 Repo Integration

To ensure separation between the Caravel files and the user project files, Caravel is included as a submodule. The caravel files can be found in the caravel folder, that folder is generated after the setup and the commit should point to the latest of caravel/caravel-lite master/main branch. Symbolic links should be established for the following files, linking them to their corresponding files in Caravel:

- **Openlane Makefile**: The Makefile simplifies the execution of Openlane for the purpose of hardening the user macros. It also retains the Openlane summary reports within the signoff directory.
  To initiate the hardening process for a user macro, execute the console make <folder_name> command. It is important to ensure that the design files of the macro are located inside the "openlane" directory.
- **Pin order file for the user wrapper**: The pin order specified in the hardened user project wrapper macro must match the pin order defined in Caravel's repository. Failure to adhere to the same order may result in the integration of the macro with Caravel's back-end failing during the GDS integration process.

7.5 Verilog Integration

In order to integrate your macro seamlessly, it is necessary to create a wrapper around it following the template provided in the verilog/rtl/ directory. The wrapper's top module should be named "user_project_wrapper" and must have identical input and output ports as specified in the golden wrapper template. This wrapper facilitates access to various user space utilities offered by Caravel, including IO ports, logic analyzer probes, and a wishbone bus connection to the management SoC. The Verilog file template for the user_project_wrapper is provided in [66].

7.6 GPIO Configuration

To ensure the proper initialization and configuration of each GPIO in Caravel, it is necessary to specify the power-on default settings for each GPIO. These default settings determine the initial state of the GPIO upon power-up. It is important to note that these configurations can be modified by the management SoC during the execution of firmware.

The configuration settings define various aspects of the GPIO, including whether it is connected to the user project area or the management SoC, the direction of the IO (input or output), whether it is digital or analog, and whether pull-up or pull-down resistors are enabled for inputs.

To customize the GPIO configurations, you can assign predefined values for each IO in the file user_defines.v within your project. This file allows you to specify the desired settings for each GPIO based on your specific requirements. By modifying the values in this file, you can ensure that the GPIOs are configured appropriately according to your project's needs.

7.7 Layout Integration

The Caravel layout is pre-designed to include an empty golden wrapper in the user space. In order to complete the integration process and prepare for tapeout, you are only required to provide a valid user_project_wrapper GDS file. During the tapeout process, your hardened user_project_wrapper will be inserted into a standard Caravel layout, resulting in the final layout that will be sent for fabrication. Figure 7.3 illustrates the integration of the caravel with the user project.

To ensure seamless integration and avoid any potential DRC or LVS issues, it is important that your hardened user_project_wrapper adheres to a set of specific requirements outlined in the User Project Wrapper Requirements. These requirements serve as guidelines to ensure compatibility and successful integration of your design with the Caravel layout. By following these requirements, you can help ensure a smooth fabrication process and achieve the desired functionality of your project within the Caravel framework.

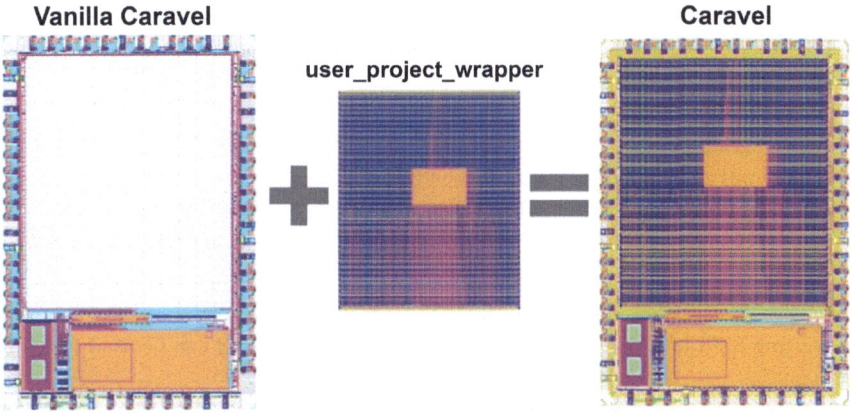

Fig. 7.3 Illustration of the Integration Process – Combining the Caravel Template with the User_Project to Form a Custom SoC [62].

7.8 RTL to GDSII Flow with Caravel

The aim of this chapter is to familiarize readers with the RTL to GDSII Flow using Caravel, a crucial step in the journey of a design from conception to realization. RTL represents the initial conceptual stage of the design, which is then materialized into a physical layout in the GDSII format. We will explore how this transition is facilitated by the Caravel harness, a vital part of the open-source EDA tool flow.

RTL designs illustrate the logic of a circuit through a high-level descriptive language, most commonly Verilog or VHDL. This RTL design is then synthesized into gate-level representations, and finally, these are converted into physical layouts for silicon fabrication, a process encapsulated in GDSII files.

The Caravel user-project wrapper, provided by Efabless, plays a pivotal role in this workflow. It offers a pre-verified environment comprising a RISC-V microcontroller unit, which allows users to seamlessly integrate their projects as hardened macros or user project areas. This enables verification and packaging of the design for fabrication using the open-source process offered by SkyWater Technology.

In this chapter, we shall deep dive into each stage involved in the RTL to GDSII process and observe how they are handled by the OpenLane tool. We will discuss how RTL code is input into the system, delve into the steps for synthesizing the design to gate-level, understand placement and routing, and finally, discuss the creation of the GDSII file.

Here are the critical guidelines you need to follow during this process:

- The top-level module must be named "user_project_wrapper". This serves as the principal module encompassing all the other modules and components.

- The "user_project_wrapper" module must adhere to the pin order defined in the "Digital Wrapper Pin Order" documentation. This promotes uniformity and interoperability with other components.
- The design configurations for the "user_project_wrapper" need to align with the "Digital Wrapper Fixed Configuration". This establishes consistent design parameters across different implementations.
- The user project repository must be organized following the "Required Directory Structure". This enhances repository management and accessibility for collaborators.

Caravel uses OpenLane to run the RTL to GDSII workflow. This process helps in converting high-level logical representations into a format that can be used for chip fabrication.

There are two common strategies to harden the user project area:

1. **Harden the user macro(s)** first, then integrate them into the "user_project_wrapper". This approach ensures that each individual component is robust before they are combined.
2. **Flatten the user macro(s)** with the "user_project_wrapper". This method merges all hierarchical layers of the design into a single layer, potentially enhancing overall performance and simplifying the design process.

7.8.1 Harden the User Macro

The initial stage in the RTL to GDSII process of a macro involves the creation of the OpenLane project. This should be done within the dedicated openlane directory.

As discussed in the previous section, OpenLane operates using a config.json file, which serves as the configuration file for the project.

Thus, as part of setting up your OpenLane project, you need to create the config.json file inside your project folder. This file will hold key configuration settings and parameters for your design, helping guide the RTL to GDSII flow.

Remember, the config.json file plays a crucial role in defining the configuration of your project, so it's important to set it up correctly. It specifies the rules and constraints for the design synthesis and physical design steps of your project, ensuring smooth and efficient execution of the RTL to GDSII process. Figure 7.4 illustrates the path and files that are indispensable to initiate the RTL to GDSII workflow.

To initiate this process systematically, it's essential to first set the environment variable - PDK . These variables act as references to critical paths within the system, allowing for smoother navigation and execution of tasks. Establishing these variables is the foundational step to ensuring the effective operation of Caravel.

To set the environment variables run:

```
project_name is the name of your repo
$cd <project_name>
```

7.8 RTL to GDSII Flow with Caravel

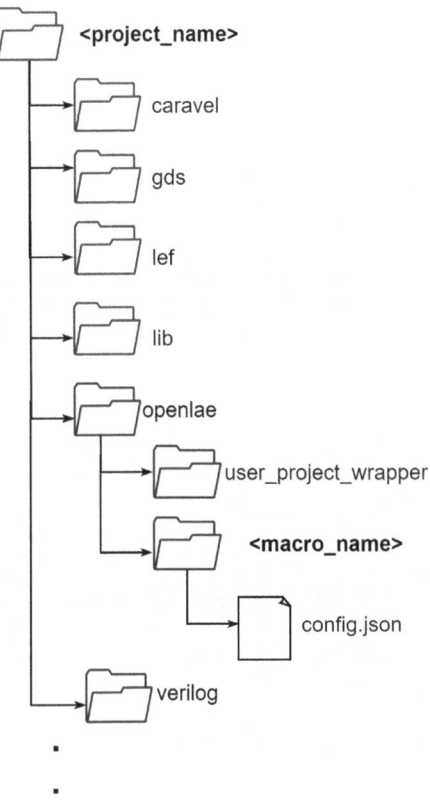

Fig. 7.4 A schematic representing the essential paths and files for launching the RTL to GDSII workflow.

```
for sky130 MPW shuttles....
$ export PDK=sky130A
```

The following step is the creation of the OpenLane project inside of the openlane folder and then creating the config.json file.

```
cd (change directory) to openlane
$ cd openlane
Create a new directory
$ mkdir <my_directory>
Navigate into the new directory
$ cd <my_directory>
Create a new .json file using
$ touch config.json
```

For example:

```
$ cd openlane
$ mkdir Counter_9999
$ cd Counter_9999
$ touch config.json
```

The subsequent stage involves the creation and verification of the Verilog files. To maintain the required Directory Structure, all Verilog files should be housed within the designated folder, specifically under the path .../verilog/rtl/ , as illustrated in Figure 7.4.

In order to illustrate the RTL to GDSII workflow, we will employ the use of a 9999 counter as an example, the verilog code can be found at [67]. Given that OpenLane is utilized for this conversion process, the steps will bear similarity across different projects. However, as illustrated in Figure 7.4, the primary difference lies in the adherence to the Directory Structure. This structure, imposed by Caravel is mandatory.

The configuration file for the 9999 counters is provided below:

```
{"DESIGN_NAME": "Counter_9999",
"DESIGN_IS_CORE": 0,
"VERILOG_FILES": [
"dir::../../verilog/rtl/defines.v",
    "dir::../../verilog/rtl/Counter_9999.v"],
"CLOCK_PERIOD": 25,
"CLOCK_PORT": "clk",
"FP_SIZING": "absolute",
"DIE_AREA": "0 0 80 80",
"VDD_NETS": "vccd1",
"GND_NETS":"vssd1",

"RT_MAX_LAYER": "met4",

"FP_PDN_HPITCH":10,
"FP_PDN_VPITCH":15,
"FP_PDN_HOFFSET":2,
"FP_PDN_VOFFSET":2,
"FP_IO_HLENGTH":2,
"FP_IO_VLENGTH":2,

"PL_RANDOM_GLB_PLACEMENT":1,
"PL_TARGET_DENSITY":0.8,
"PL_BASIC_PLACEMENT":1}
```

7.8 RTL to GDSII Flow with Caravel

In light of previously mentioned parameters that can be found in the OpenLane documentation, our focus will be trained on the crucial differences and guidelines necessary for successful SoC integration. First, it's important to note that the design is treated as a macro, hence the DESIGN_IS_CORE option is set to 0. As this module operates as a macro, we are required to define the voltage and ground pins with the VDD_NETS and GND_NETS variables respectively.

One key aspect is the setting of the RT_MAX_LAYER option. While it's possible to set it to a layer higher than met4, it's necessary to note that during the final stages of the flow, the user project area will be integrated with the Caravel management area which utilizes the other layers for routing the entire SoC. Therefore, setting the layer properly plays a crucial role in the smooth integration of the design, so the maximum routing layer will be met4.

Another point of emphasis is the VERILOG_FILES variable. This variable should be directed to the verilog/rtl/ path (see Figure 7.4 for references), reinforcing the importance of maintaining the directory structure discussed earlier.

To initiate the RTL to GDSII conversion process, simply input the command detailed below:

```
$ make <module_name>
e.g.
$ make Counter_9999
```

If the process executes correctly, you will encounter the message "[SUCCESS]: Flow complete." However, if any issues arise during the execution, the system will output log messages detailing the problem. In such a case, it is necessary to scrutinize these messages to identify the error and proceed with the appropriate corrections.

The subsequent step involves integrating our module with the "user_project _wrapper" file, effectively meaning that we need to instantiate our module within the "user_project_wrapper" module and establish the necessary connections. The instantiation and connections can be found at [68].

Proceed to configure the user_project_wrapper/config.json file accordingly, which can be found at [69]. This involves setting up parameters to match the specifications of your project. It's vital to ensure that all entries align with your design requirements for a successful RTL to GDSII workflow execution.

Lastly, execute the RTL to GDSII workflow for the user_project_wrapper. Please ensure you update the necessary parameters in the user_project_wrapper/config.json file, such as CLOCK_NET, FP_PDN_MACRO_HOOKS, VERILOG_FILES _BLACKBOX, EXTRA_LEFS (located in the lef folder), EXTRA_GDS_FILES (found in the gds folder), and set up the macro.cfg file appropriately.

Figure 7.5 showcases the GDSII file following the hardening process of the user_project_wrapper, utilizing the 9999 counter as a macro.

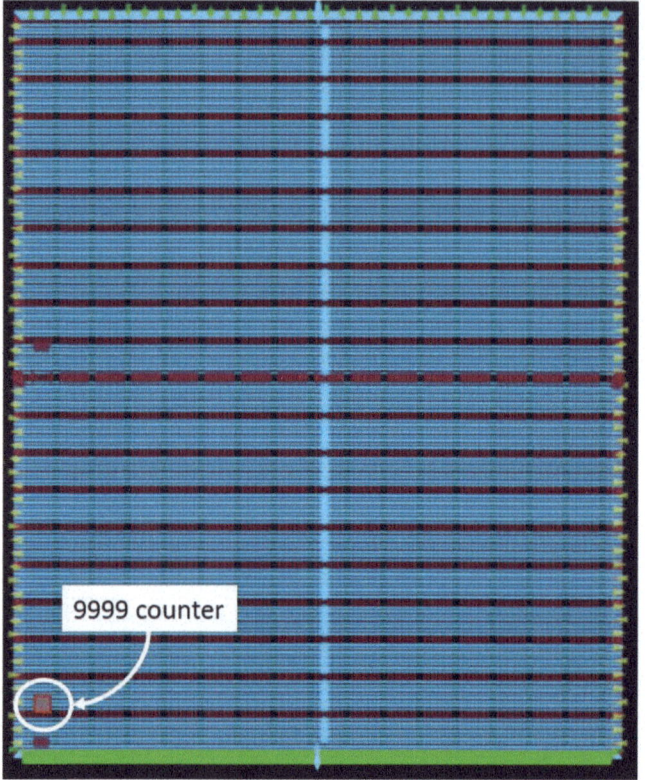

Fig. 7.5 GDSII file of user_project_wrapper post-hardening, with the 9999 counter as macro.

The interconnections between the user project area and the management area will be examined in detail in the subsequent sections.

7.8.2 LVS Issues with Voltage and Ground Pins

The mismatch of VCC and VSS is a frequent issue that arises when utilizing this methodology for RTL to GDSII flow. To ensure that an LVS error pertains to the VCC and VSS mismatch, inspect the .lvs.log file situated in the path user_project_wrapper/runs/<tag>/logs/ , note that the path structure in the open-Lane folder is the same as that presented in Figure 4.2. The log should display a no matching pin message in the Subcircuit summary: section.

If this section shows a non-matching VCC and VSS message, it implies that your module is not connected to the power grid. To remedy this, you can alter the size of the power grid and the spacing between each power line. Another solution involves changing the placement of the module by modifying the macro.cfg file.

7.8 RTL to GDSII Flow with Caravel

Figure 7.6 illustrates the placement of the Counter_9999 module both with and without a VCC and VSS mismatch. By comparing these scenarios, you can gain a clear understanding of how the module's connection to the power grid impacts its functionality and how necessary adjustments can resolve the VCC and VSS mismatch problem.

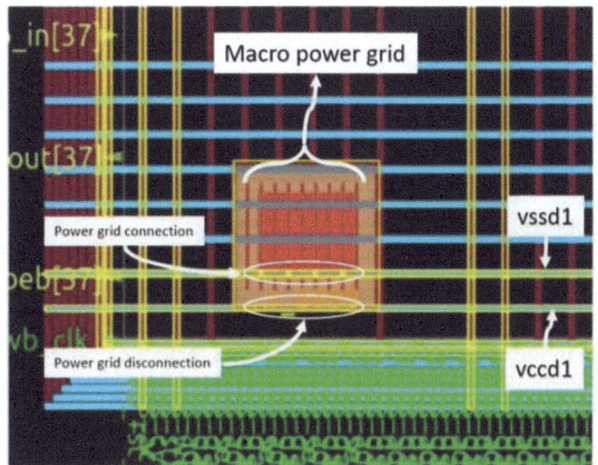

(a) Mismatch of VCC and VSS

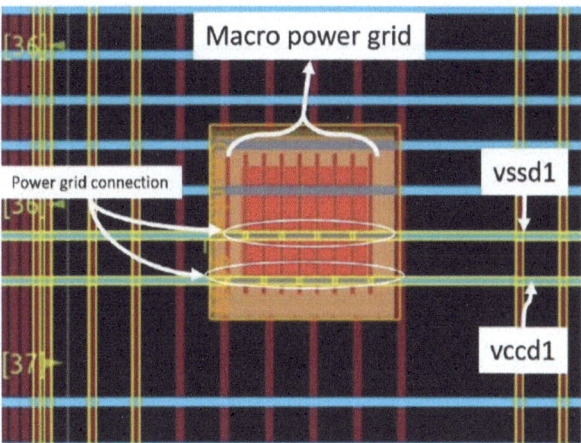

(b) Matching of VCC and VSS

Fig. 7.6 Comparison of "Counter_9999" Module Placement with VCC and VSS Connections. (a) Depicts the system with a VCC and VSS mismatch, demonstrating an instance where the module is not correctly connected to the power grid. (b) Shows the system without the VCC and VSS mismatch, highlighting the corrected module placement and proper connections to the power grid.

7.8.3 Flatten the User Macro

The process explained in the previous section emphasizes a systematic initiation, which importantly involves setting key environment variable PDK. Setting this variable forms the bedrock of efficient Caravel operation. To configure these environment variables, execute the following command:

```
project_name is the name of your repo
$cd <project_name>
for sky130 MPW shuttles....
$ export PDK=sky130A
            or
for the gf180 GFMPW shuttles...
$ export PDK=gf180mcuC
```

The main difference between both methods lies in their setup process. In the macro hardening method, it is necessary to create a macro along with all its configuration files. However, in the Flatten method, you simply create the desired Verilog file and instantiate it within the user_project_wrapper module. As an example, the Verilog files for the Counter_9999 can be found at [70].

The key difference to note with this module is that the VCC and VDD ports are not necessary for this module declaration. The user_project_wrapper module will now appear as [71] illustrate. Take note that the Counter_9999 module does not include VCC and VSS pins. The configuration variables needed to harden this module can be found at [72].

Significant changes are reflected in the variables required for a core implementation with macros. In this case, since the user_project_wrapper is considered as a unified system, those specific macro-related variables are not necessary. Figure 7.7 showcases the GDSII file following the hardening process of the user_project_wrapper, flattening the 9999.

7.9 SoC Integration

In this critical chapter, we will delve into SoC integration, focusing specifically on the connection between the user project area and the management area. Understanding this process is essential for creating a successful, unified design that functions efficiently and effectively.

A key part of the Caravel project, the SoC brings together multiple discrete components into one singular, cohesive system. This system, as we have already discussed, includes the VexRiscv core, memory, a flash controller, and serial peripherals within the management area. Additionally, it also includes the user project area, which can contain various user-defined designs and macros.

7.9 SoC Integration

Fig. 7.7 GDSII file of user_project_wrapper post-hardening, with the 9999 counter as a flat design

Integration of these areas allows for interaction and communication between user-defined projects and the standardized management SoC, creating a functional, custom, and comprehensive digital design.

In this chapter, we will explain in detail how these different components of the SoC - user projects and the management area - are connected. We will highlight the integration process and the necessary considerations to ensure proper functionality and performance.

By understanding the process of SoC integration, you will be able to effectively combine your digital design projects with the existing components of the SoC to produce a robust and versatile chip. Whether your projects involve new processor designs, digital signal processing units, or novel digital circuits, this chapter will provide you with the knowledge and skills to integrate your projects into the larger SoC design.

7.9.1 User Area Pins

The pins for the user project area in the Caravel SoC are organized into five distinct categories for efficient management and functionality. These categories are:

1. **Wishbone**: This set of pins forms the Wishbone interface, an open standard bus protocol often used in SoC architectures for data exchange. It provides a well-defined method for exchanging data between different components, ensuring compatibility and coherence within the system.

   ```
   // Wishbone Slave ports (WB MI A)
       input wb_clk_i,
       input wb_rst_i,
       input wbs_stb_i,
       input wbs_cyc_i,
       input wbs_we_i,
       input [3:0] wbs_sel_i,
       input [31:0] wbs_dat_i,
       input [31:0] wbs_adr_i,
       output wbs_ack_o,
       output [31:0] wbs_dat_o,
   ```

2. **Logic Analyzer**: These pins are dedicated to the Logic Analyzer, a useful tool for debugging and testing. Through these pins, the Logic Analyzer can monitor signals within the SoC, providing crucial insight into the system's operations.

   ```
   // Logic Analyzer Signals
       input  [127:0] la_data_in,
       output [127:0] la_data_out,
       input  [127:0] la_oenb,
   ```

3. **General-Purpose Input/Output**: The GPIO pins allow for flexible bidirectional communication with a variety of peripherals or devices. This group of pins can be programmed to function as either input or output and handle a variety of functions depending on the user's design requirements.

   ```
   // IOs
       input  [`MPRJ_IO_PADS-1:0] io_in,
       output [`MPRJ_IO_PADS-1:0] io_out,
       output [`MPRJ_IO_PADS-1:0] io_oeb,
   ```

4. **Clock**: The clock pins provide timing signals to synchronize the operations of different components within the SoC. This is essential to maintain the order and timing of operations, ensuring that all elements of the design work together smoothly.

```
        // Independent clock (on independent integer divider)
        input   user_clock2,
                // or
        // Wishbone Slave ports (WB MI A)
        input wb_clk_i,
```

5. **Interrupts**: The interrupt pins provide a mechanism for handling events that require immediate attention by the processor. They signal the processor to temporarily pause its current operations and handle the interrupting event, which can range from peripheral requests to system-level events.

```
        // User maskable interrupt signals
        output [2:0] user_irq
```

Understanding how these different pin categories function within the user project area is vital for effective SoC integration, as they establish the interfaces between the user's digital designs and the rest of the Caravel SoC.

7.9.2 Logic Analyzer Pins

The Logic Analyzer pins provide a powerful means of observing the internal operations of the SoC design. With a total of 128 input pins and 128 output pins available, these allow for deep insight into the system's behavior. However, the ample number of pins, the management area can concurrently handle a total of only 128 pins. This limitation is due to the current architecture's design, where the switching between input and output is controlled via register configuration by the RISC-V processor.

To tackle situations where more than 128 signals need to be analyzed, a creative workaround can be employed using a bank of registers. These registers can temporarily hold the values of both input and output signals. The RISC-V processor can then manage the reading and writing of these signals to and from the registers, effectively extending the logic analysis capabilities of the system. This method, while slightly more complex, ensures that larger and more intricate designs can be adequately monitored and debugged, enhancing the robustness of the SoC's verification process.

7.9.3 GPIO Pins

Like the Logic Analyzer pins, the GPIO pins also offer a significant degree of flexibility in the system configuration. A total of 38 input pins and 38 output pins are available for user-defined tasks, the direction and function of which can be selected via the RISC-V processor or via the user project area.

As highlighted in Chapter 10, these general-purpose ports must be suitably configured to match the intended function within the SoC design. This is achieved by modifying the user_defines.v file. This file serves as the primary interface for setting the initial configuration of the GPIO pins, effectively enabling users to tailor the I/O characteristics to match their specific project requirements. If a GPIO port is configured with the MGMT_STD option, it delegates control of the port's operational direction to the RISC-V processor. This means that the RISC-V core determines whether the port serves as an input or output, dynamically adjusting based on the software running on the processor.

Conversely, if the GPIO port is configured with the USER_STD option, the responsibility of managing the port's functionality rests with the user's project. In this case, the direction of the port (input or output) is handled within the user project logic, providing more direct control over the port's behavior. This flexibility enables users to tailor the GPIO configuration closely to the specific requirements of their design. If you aim to be a part of the MPW run, it's essential to configure these GPIO pins' initial assignment correctly. This step is not optional. If not done properly, the final design validation checks will yield errors, potentially risking your inclusion in the MPW run. Thus, the correct and accurate configuration of these pins plays a crucial role in the successful execution and fabrication of your design.

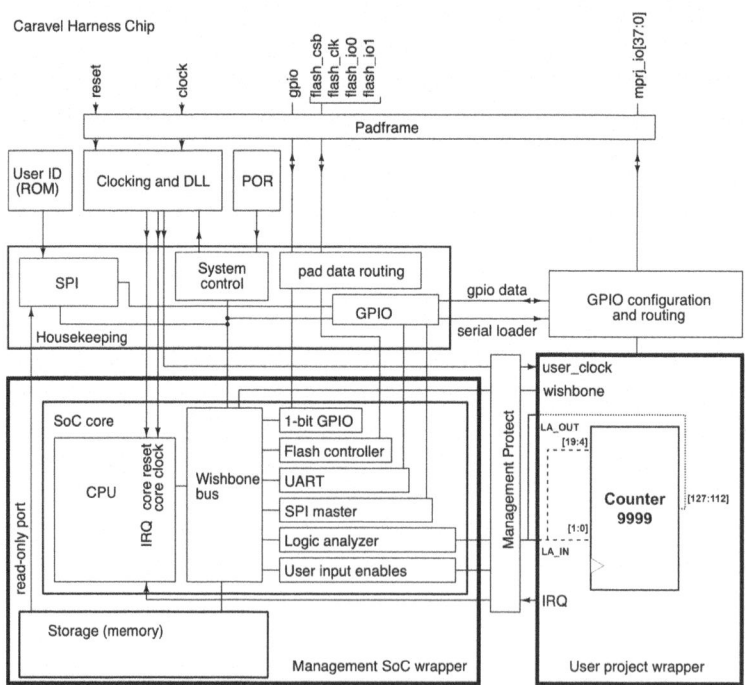

Fig. 7.8 Schematic of the System - Highlighting the interface between the RISC-V processor and the 9999 counter via the Logic Analyzer ports. The figure illustrates the distribution of bit assignments for control signals, parallel loading, and current count reading.

7.10 SoC Integration Example

For this example, we will utilize the previously seen 9999 counter. Fig 7.8 depicts the schematic of our system, wherein the RISC-V processor communicates with the counter via the Logic Analyzer ports.

We will allocate the first 4 bits, la_data_in[3:0], for control signals. The subsequent 16 bits, la_data_in[19:4], will be designated for parallel loading. The last 16 bits, la_data_out[127:112], will be utilized to read the current count. Lastly, the clock signal wb_clk_i will be used. Regardless of which hardening option you select, the above connections will remain the same. However, do remember that if you've chosen the first hardening option, you will need to connect the VCC and VSS pins. The Verilog file for the user_project_wrapper can be found at [73].

Regardless of the state of use of the GPIOs, they require initial configuration. This rule applies even if these ports are not actively utilized in your specific project. [74] shows an example of the GPIO status. Once the connections are established, the next step involves the hardening of the user_project_wrapper, executing the command.

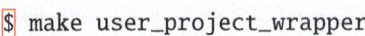
```
$ make user_project_wrapper
```

If any errors occur during the workflow, be sure to review the message logs and modify the necessary configuration parameters.

References

1. McKinsey & Company. The semiconductor decade: A trillion-dollar industry. https://www.mckinsey.com/industries/semiconductors/our-insights/the-semiconductor-decade-a-trillion-dollar-industry, 2023. Accessed: 2024-07-22.
2. Deloitte. Global semiconductor talent shortage. https://www2.deloitte.com/us/en/pages/technology/articles/global-semiconductor-talent-shortage.html, 2024. Accessed: 2024-07-23.
3. Tao Li, Jie Hou, Jinli Yan, Rulin Liu, Hui Yang, and Zhigang Sun. Chiplet heterogeneous integration technology—status and challenges. *Electronics*, 9(4):670, 2020.
4. IBS. As chip design costs skyrocket, 3 nm process node is in jeopardy. https://www.extremetech.com/computing/272096-3nm-process-node, 2020. Accessed: 2024-07-23.
5. A. Belous and V. Saladukha. Digital ic and system-on-chip design flows. In *The Art and Science of Microelectronic Circuit Design*, pages 317–353. Springer, Cham, 2022.
6. George Calhoun. The us still dominates in semiconductors: China is vulnerable (part 2). https://www.forbes.com/sites/georgecalhoun/2021/10/11/the-us-still-dominates-in-semiconductors-china-is-vulnerable-pt-2/, 2021. Accessed: 2024-07-23.
7. World Population Review. Semiconductor manufacturing by country. https://worldpopulationreview.com/country-rankings/semiconductor-manufacturing-by-country, 2024. Accessed: 2024-07-23.
8. Defense Advanced Research Projects Agency (DARPA). About darpa. https://www.darpa.mil/about-us/about-darpa. Accessed: 2024-07-30.
9. United States Congress. The proposed dwight d. eisenhower memorial. https://www.congress.gov/event/112th-congress/house-event/LC3882/text. Accessed: 2024-07-30.
10. Defense Advanced Research Projects Agency (DARPA). Darpa's stealth revolution. https://www.darpa.mil/about-us/timeline/darpas-stealth-revolution. Accessed: 2024-07-30.
11. Katie Hafner and Matthew Lyon. *Where wizards stay up late: The origins of the Internet*. Simon and Schuster, 1998.
12. IEEE Spectrum. Darpa's planning a major remake of us electronics: Pay attention. https://spectrum.ieee.org/darpas-planning-a-major-remake-of-us-electronics-pay-attention. Accessed: 2024-07-30.
13. Defense Advanced Research Projects Agency (DARPA). Intelligent design of electronic assets. https://www.darpa.mil/program/intelligent-design-of-electronic-assets. Accessed: 2024-07-30.
14. SR Cray and RN Kisch. A progress report on computer applications in computer design. In *Papers presented at the February 7-9, 1956, joint ACM-AIEE-IRE western computer conference*, pages 82–85, 1956.
15. Ivan E Sutherland. Sketchpad: A man-machine graphical communication system. In *Proceedings of the May 21-23, 1963, spring joint computer conference*, pages 329–346, 1963.
16. Laurence W. Nagel and D.O. Pederson. Spice (simulation program with integrated circuit emphasis). Technical Report UCB/ERL M382, EECS Department, University of California, Berkeley, Apr 1973.
17. John K. Ousterhout, Gordon T. Hamachi, Robert N. Mayo, Walter S. Scott, and George S. Taylor. A collection of papers on magic. Technical Report UCB/CSD-83-154, EECS Department, University of California, Berkeley, Dec 1983.
18. Mario R Barbacci. A comparison of register transfer languages for describing computers and digital systems. *IEEE Transactions on Computers*, 100(2):137–150, 1975.

References

19. Mario R Barbacci. Instruction set processor specifications (isps): The notation and its applications. *IEEE Transactions on Computers*, 100(1):24–40, 1981.
20. Guglielmo Girardi, Reiner W Hartenstein, and Udo Welters. Karl (textual) and abl (graphic): A user/designer interface in microelectronics. *CAD-Schnittstellen und Datentransferformate im Elektronik-Bereich*, pages 1–12, 1987.
21. Moe Shahdad, Roger Lipsett, Erich Marschner, Kellye Sheehan, and Howard Cohen. Vhsic hardware description language. *Computer*, 18(2):94–103, 1985.
22. IEEE Design Automation Sub-Committee et al. Ieee standard hardware description language based on the verilog(r) hardware description language. *IEEE Std 1364-1995*, pages 1–688, 1996.
23. ANSYS, Inc. ANSYS. https://www.ansys.com/, Accessed 2024-05-14.
24. Cadence Design Systems. Cadence. https://www.cadence.com/en_US/home.html, Accessed 2024-05-14.
25. Synopsys, Inc. Synopsys. https://www.synopsys.com/, Accessed 2024-05-14.
26. Mentor Graphics, a Siemens Business. Mentor graphics. https://eda.sw.siemens.com/en-US/, Accessed 2024-05-14.
27. The OpenROAD Project. Openroad: Democratizing hardware design. https://theopenroadproject.org/. Accessed: 2024-05-14.
28. The OpenLane Project. Openlane: An autonomous rtl-gdsii flow for vlsi designs. https://openlane.readthedocs.io/en/latest/. Accessed: 2024-05-14.
29. Ricardo Augusto Da Luz Reis. Eda: Overview and some trends. *Journal of Integrated Circuits and Systems*, 17(3):1–10, 2022.
30. Alberto Sangiovanni-Vincentelli. The tides of eda. *IEEE Design & Test of Computers*, 20(6):59–75, 2003.
31. Andreas Kuehlmann. *The best of ICCAD: 20 years of excellence in computer-aided design*. Springer Science & Business Media, 2012.
32. Louis Scheffer and Luciano Lavagno. *EDA for IC System Design, Verification, and Testing*. CRC press, 2006. This book provides an overview of EDA tools and methods used in the design, verification, and testing of integrated circuits.
33. Khosrow Golshan. *Physical design essentials*. Springer, 2007.
34. Skywater pdk - github repository. https://github.com/google/skywater-pdk. Accessed: 2024-05-28.
35. Skywater pdk documentation. https://skywater-pdk.readthedocs.io/en/main/index.html. Accessed: 2024-05-28.
36. Gf180mcu pdk - github repository. https://github.com/google/gf180mcu-pdk. Accessed: 2024-05-28.
37. Gf180mcu pdk documentation. https://gf180mcu-pdk.readthedocs.io/en/latest/index.html. Accessed: 2024-05-28.
38. Mohamed Shalan and Tim Edwards. Building openlane: a 130nm openroad-based tapeout-proven flow. In *Proceedings of the 39th International Conference on Computer-Aided Design*, pages 1–6, 2020.
39. Denis Zezin. Modern open source ic design tools for electronics engineer education. In *2022 VI International Conference on Information Technologies in Engineering Education (Inforino)*, pages 1–4. IEEE, 2022.
40. S Charaan, S Nalinkumar, P Elavarasan, P Prakash, and P Kasthuri. Design of an all-digital phase-locked loop in a 130nm cmos process using open-source tools. In *2022 International Conference on Electronic Systems and Intelligent Computing (ICESIC)*, pages 270–274. IEEE, 2022.
41. Ahmed Ghazy and Mohamed Shalan. Openlane: The open-source digital asic implementation flow. In *Proc. Workshop on Open-Source EDA Technol.(WOSET)*, 2020.
42. Mikhail Chupilko, Alexander Kamkin, and Sergey Smolov. Survey of open-source flows for digital hardware design. In *2021 Ivannikov Memorial Workshop (IVMEM)*, pages 11–16. IEEE, 2021.

43. Sarah Hesham, Mohamed Shalan, M Watheq El-Kharashi, and Mohamed Dessouky. Digital asic implementation of risc-v: Openlane and commercial approaches in comparison. In *2021 IEEE International Midwest Symposium on Circuits and Systems (MWSCAS)*, pages 498–502. IEEE, 2021.
44. OpenLane. Configuration. https://openlane.readthedocs.io/en/latest/reference/configuration.html, 2024. Accessed: 2024-05-24.
45. Docker. Install docker desktop on linux. https://docs.docker.com/desktop/install/linux-install/, 2024. Accessed: 2024-05-24.
46. David Harris and Sarah Harris. *Digital design and computer architecture*. Morgan Kaufmann, 2010.
47. Cem Ünsalan and Bora Tar. *Digital system design with FPGA: Implementation using Verilog and VHDL*. McGraw Hill, 2017.
48. Thomas L Floyd. *Digital fundamentals, 10/e*. Pearson Education India, 2011.
49. Jayaram Bhasker and Rakesh Chadha. *Static timing analysis for nanometer designs: A practical approach*. Springer Science & Business Media, 2009.
50. Emilio Baungarten. Verilog code 16-bit counter. https://github.com/Baungarten-CINVESTAV/Tape-out-process-with-Open-Source-Tools/blob/main/Chapter%204/_16bit_counter.v, 2024. Accessed: 2024-07-03.
51. Majid Sarrafzadeh, Maogang Wang, and Xiaojian Yang. *Modern placement techniques*. Springer Science & Business Media, 2003.
52. Emilio Baungarten. Verilog alu_8bit. https://github.com/Baungarten-CINVESTAV/Tape-out-process-with-Open-Source-Tools/blob/main/Chapter%206/ALU_8Bit.v, 2024. Accessed: 2024-07-03.
53. Neil HE Weste and David Harris. *CMOS VLSI design: a circuits and systems perspective*. Pearson Education India, 2015.
54. Emilio Baungarten. Verilog sky130_sram. https://github.com/Baungarten-CINVESTAV/Tape-out-process-with-Open-Source-Tools/blob/main/Chapter%206/sky130_sram_1kbyte_1rw1r_8x1024_8.v, 2024. Accessed: 2024-07-03.
55. Emilio Baungarten. Verilog fifo. https://github.com/Baungarten-CINVESTAV/Tape-out-process-with-Open-Source-Tools/blob/main/Chapter%206/FIFO.v, 2024. Accessed: 2024-07-03.
56. Francesco Dell' Anna. Non linear pseudo random generator, 2020. Accessed: 2024-10-14.
57. david lundgren. Double-precision floating point unit, 2019. Accessed: 2024-10-14.
58. Richard Herveille. I2c controller core, 2018. Accessed: 2024-10-14.
59. Javier Castillo Villar. 128/192 aes, 2019. Accessed: 2024-10-22.
60. Abdul Muheet Ghani. Rv32i single cycle processor, 2024.
61. Efabless Corporation. Caravel: A template SoC for Open-Source ASIC exploration. https://github.com/efabless/caravel, 2023. [Online; accessed 18-July-2023].
62. Efabless Corporation. Caravel User Project - Caravel Harness documentation. https://caravel-harness.readthedocs.io/en/latest/, 2021.
63. Efabless Corporation. Caravel user project. https://caravel-user-project.readthedocs.io/en/latest/, 2020. Accessed: 2024-06-03.
64. Kassem Mohamed, et al. Efabless. https://efabless.com/, Dec 2020.
65. Efabless Corporation and SkyWater Technology. Chipignite 2204C. https://efabless.com/chipignite, May 2021.
66. Emilio Baungarten. Verilog user_project_wrapper. https://github.com/Baungarten-CINVESTAV/Tape-out-process-with-Open-Source-Tools/blob/main/Chapter%207/user_project_wrapper.v, 2024. Accessed: 2024-07-03.
67. Emilio Baungarten. Verilog counter_9999. https://github.com/Baungarten-CINVESTAV/Tape-out-process-with-Open-Source-Tools/blob/main/Chapter%207/Counter_9999.v, 2024. Accessed: 2024-07-03.
68. Emilio Baungarten. Verilog user_project_integration. https://github.com/Baungarten-CINVESTAV/Tape-out-process-with-Open-Source-Tools/blob/main/Chapter%207/user_project_integration.v, 2024. Accessed: 2024-07-03.

References

69. Emilio Baungarten. Json config_user_project. https://github.com/Baungarten-CINVESTAV/Tape-out-process-with-Open-Source-Tools/blob/main/Chapter%207/config_user_project.json, 2024. Accessed: 2024-07-03.
70. Emilio Baungarten. Verilog counter_9999_flatten. https://github.com/Baungarten-CINVESTAV/Tape-out-process-with-Open-Source-Tools/blob/main/Chapter%207/Counter_9999_Flatten.v, 2024. Accessed: 2024-07-03.
71. Emilio Baungarten. Verilog user_project_flatten. https://github.com/Baungarten-CINVESTAV/Tape-out-process-with-Open-Source-Tools/blob/main/Chapter%207/user_project_wrapper_Flatten.v, 2024. Accessed: 2024-07-03.
72. Emilio Baungarten. Json config_user_flatten. https://github.com/Baungarten-CINVESTAV/Tape-out-process-with-Open-Source-Tools/blob/main/Chapter%207/config_user_project_Flatten.json, 2024. Accessed: 2024-07-03.
73. Emilio Baungarten. Verilog soc_integration_example. https://github.com/Baungarten-CINVESTAV/Tape-out-process-with-Open-Source-Tools/blob/main/Chapter%207/SoC_Integration_Example.v, 2024. Accessed: 2024-07-03.
74. Emilio Baungarten. Verilog gpio. https://github.com/Baungarten-CINVESTAV/Tape-out-process-with-Open-Source-Tools/blob/main/Chapter%207/GPIO.v, 2024. Accessed: 2024-07-03.

The manufacturer's authorised representative in the EU is Springer Nature Customer Service Centre GmbH, Europaplatz 3, 69115 Heidelberg, Germany. If you have any concerns regarding our products, please contact ProductSafety@springernature.com

Printed and bound by CPI Group (UK) Ltd, Croydon, CR0 4YY

26/03/2026

02078943-0014